TRANSFORMATIVE ADAPTATION

Rupert Read and Morgan Phillips,
with Manda Scott

for and with the TrAd collective

Permanent Publications

Published by
Permanent Publications
Hyden House Ltd
13 Clovelly Road
Portsmouth
PO4 8DL
United Kingdom
Tel: 01730 776 582
Email: enquiries@permaculture.co.uk
Web: www.permanentpublications.co.uk

Distributed in Australia by
Peribo Pty Limited, 58 Beaumont Road, Mt Kuring-Gai, NSW 2080 Australia
https://peribo.com.au

© 2024 Rupert Read and Morgan Phillips

The right of Rupert Read and Morgan Phillips to be identified as the authors of this work has been asserted by them in accordance with the Copyrights, Designs and Patents Act 1998

Cover Design by Two Plus George Limited, info@twoplusgeorge.co.uk

Designed by Rozie Apps, Permanent Publications

Printed in the UK by Bell & Bain, Thornliebank, Glasgow

MIX
Paper | Supporting responsible forestry
FSC® C007785

This product is made of material from well-managed FSC®-certified forests and from recycled materials and other controlled sources.

The Forest Stewardship Council ® (FSC) is a non-profit international organisation established to promote the responsible management of the world's forests. Products carrying the FSC label are independently certified to assure consumers that they come from forests that are managed to meet the social, economic and ecological needs of present and future generations.

British Library Cataloguing-in-Publication Data
A catalogue record for this book is available from the British Library

ISBN 978 1 85623 225 8

All rights reserved. No part of this publication may be reproduced, stored in a retrieval system, rebound or transmitted in any form or by any means, electronic, mechanical, photocopying, recording or otherwise, without the prior permission of Hyden House Limited.

TRANSFORMATIVE ADAPTATION

The first word on an emerging response to the climate and ecological more-than-emergency.

It is absolutely essential that the oppressed participate in the revolutionary process with an increasingly critical awareness of their role as subjects of the transformation.

Paulo Freire

Praise for the book

The same global economic system that is responsible for ecocide is impoverishing the middle class and driving them into the arms of demagogues. We need a big picture to unite a people's movement, focused on bringing about fundamental systems change. I am hopeful that the essays in this book contribute to drawing exactly that picture.

Helena Norberg-Hodge,
Founder and Director of Local Futures

Often witty, always wise, this book artfully articulates what the love and respect for our home and one another that we say we hold dear, might actually look like if we really lived it. The subsequent expansion of imaginative possibilities is authentically inspiring, creating a compelling call for change, together.

Ed Gillespie,
Futurenaut

If we are to have any chance of surviving the prehistoric climate that global heating is certain to bring, we need a complete transformation of how society and economy function and how we live our lives. It is a colossal ask, but we can do it. In this marvellous and inspiring book, Morgan Phillips, Rupert Read, and others, paint an optimistic picture of what this transformation will look like.

Bill McGuire,
Professor Emeritus of Geophysical & Climate Hazards at UCL,
and author of *Hothouse Earth: an Inhabitant's Guide*

We know the collapse is coming, and that our attempts to avert the collapse by demanding change have not worked. It is clear that those directing the economic nonsense cannot avoid the collapse they have been orchestrating, nor are they willing to take action to cushion the fall. This hopeful and inspiring book sets us on the path of adapting to the collapse in real time, and simultaneously creating ways of being, doing and organising that clear a path towards a new civilisation.

Diana Finch,
author of *Value Beyond Money*

Beyond awakening, this book is a call to action. Right here and now we can each step fully into the chrysalis of civilisational decline, and we can sense into what new ways of being and doing want to emerge through us. This book gives us glimpses into what is possible, if we align with nature, come together, ignite our imagination, and take collective action. Read this book. Find what is yours to do. And together let's create our thrutopias!

Jessica Bockler,
Director of Alef Trust

About the Authors

Rupert Read is co-director of the Climate Majority Project and Emeritus Professor of Philosophy at the University of East Anglia. He is the author of several books, developed the concept of Transformative Adaptation and also invented the term 'Thrutopia'; an act of imagining the actual mechanics of how we transition from this broken, collapsing world to one that is in a process of transformation and healing on all levels. Since then he has been building Thrutopianism into his creative and scenarios work, including his viral internet short *Out of the Ashes*: www.youtube.com/watch?v=vi166hJv6Qk.

https://rupertread.net
https://climatemajorityproject.com

Morgan Phillips is Global Action Plan's Director of Education and Youth Engagement, and a former co-director of climate change adaptation charity, The Glacier Trust, with whom he continues to volunteer. Morgan has worked in the environmental sector for over 20 years in roles that have taken him to Nepal, Kenya, Jamaica, Bangladesh and Slovakia. Having returned home to west Wales in 2022, Morgan is now governor at his local primary school and a member of the education committee at Black Mountains College. Morgan's 2021 book *Great Adaptations – In the shadow of a climate crisis* was published by Arkbound UK and has since been published in Japanese.

Skeena Rathor is a trauma therapist, body and brain growth mentor, heartfulness, and heartmath teacher. She was the Co-founder of the Guardianship and Visioning team of Extinction Rebellion (XR). She is a leader in 'Being the Change Affinity Network'. She has frequently been on TV and radio.

Suzanne O'Hara is a fellow of the School for Social Entrepreneurs and Co-Founding Organiser at Talking Tree climate emergency hub. She is a food and drinks marketing veteran, adviser to purpose-led businesses and has a special interest in food system initiatives to improve the health of people and planet.

Ben McCallan is the former chair of charity Zero Carbon Guildford, a project focusing on driving climate and environment action across a wide demographic of residents, as well as the Engagement Lead across climate change, strategic energy and natural environment at Surrey County Council.

Rachel Bailey is a mother of two children and a Producer of media and events across education and mental health. She was part of XR's Visioning Sensing team, and several other groups, working towards bringing forth new stories for our shared future.

Ruth Allen is co-creator of the Climate Emergency Centre network and CCPAST (Climate Child Protection and Safeguarding Team). She is a social worker and therapist, and mother to two amazing children, and two hairy dogs.

A founder of the Community Benefit Society iFarm (10 acres of land in Norfolk), that has also taken on the local pub (The White Horse) as a hub for activity, Joolz Thompson has helped to bring a community together to deliver Community Climate Action planning (an initiative funded by the National Lottery and incubated by the Climate Majority Project). Joolz is a Parish Councillor in Hopton cum Knettishall and works for Suffolk Farming and Wildlife Group, as a Business Development Consultant – helping farmers transition to more sustainable agriculture and connecting communities with those who grow their food. Joolz is Non-executive Director of The Great Collaboration(a toolkit designed to support Council and Community Climate Action).

Formerly a veterinary surgeon, Manda Scott is now an award-winning novelist, podcaster and regenerative smallholder. Co-creator of the Thrutopia Masterclass, she is committed to laying the foundations to a future we'd be proud to leave behind. **https://accidentalgods.life**.

Contents

Preface: From Disgrace to Grace Rupert Read		1
Foreword Chris Smaje		5
1.	**A Moment of Criticality** Rupert Read	9
2.	**The Age of Adaptation** Morgan Phillips and Rupert Read	13
3.	**Transformative Adaptation** Rupert Read	17
4.	**How we will Free Ourselves – Together** Skeena Rathor and Rupert Read	25
5.	**Agroforestry Hubs** **– Local Models of Transformative Adaptation** Morgan Phillips	29
6.	**Climate Hubs – A Transformative Way** Rachel Bailey, Ruth Allen, Ben McCallan and Suzanne O'Hara	37
7.	**Transformative Adaptation as Part of** **the Emerging Climate Majority** Rupert Read	45
8.	**TrAd and Community Climate Action** Joolz Thompson	51
9.	**I Don't Want You to be 'More Like Me',** **I Want You to be 'More Like You'** Morgan Phillips	57
10.	**Thrutopia: Creating a New Story for** **a World Undergoing Transformation** Manda Scott	63
11.	**To TrAd or not to TrAd?** Morgan Phillips	91
Transformative Adaptation: A Declaration The TrAd Collective		95
Endnotes		98

Preface

From Disgrace to Grace

By Rupert Read

LAND: We are nothing without Earth. The land is ours – to tend and regeneratively harvest. This is non-negotiable; we will not accept obstacles placed in the way of this loving, literally-nourishing relationship.

COMMUNITY: We are nothing without each other. The future will be one of more successful, nourishing, local workings together; or it will not *be*.

TRANSFORMATION: We will be reduced to nothing unless our adaptation to this climate-damaged world is transformative. We therefore will change society in the direction it needs changing in any way, to make life better: more secure, more hands-on, more beautiful.

The three short paragraphs above are a succinct shared summation of the ethos, the 'three pillars', of Transformative Adaptation (TrAd); like a number of key elements of this book, they were devised together by the TrAd Collective.[1]

The struggle to define adaptation will, in our view, be nothing less than the defining struggle of the next decade or more. This book represents our endeavour thus far in that struggle.

The mystery is not why this is needed, nor why we are here. The mystery is why there has not been far more attention – including of the kind we are offering – to adaptation *already*. The mystery is why there has been such resistance to adaptation-narratives and -practices... given that, unlike what is called in the lingo 'mitigation' of the climate crisis (i.e. direct efforts to reduce greenhouse-gas emissions, etc.), adaptation offers immediate localised 'co-benefits'. The core of the answer that our book offers is that, when we realise that we need to talk about adaptation, *then we can no longer pretend that everything is going to be fine*. Continuing *that* pretence has, up to this point, been something that most 'environmentalists' have typically done just as much as most politicians (not to mention climate-deniers) have.

The obverse, the corollary, of the point I've just made is hugely exciting, and points to a potential significance of this book far beyond its slim pages: adaptation of the right kind *wakes people up* like nothing else. It is a slow, endless mind-bomb. When people hear talk of 2050 or 2100, in relation to talk of 'reducing emissions', too often their eyes glaze over. But when people see their neighbours or compatriots (or indeed 'the authorities') engaging in practical forms of adaptation, they *get* it. Human-triggered climate decline is real. It's coming. It's here. These people — us, people — are actually doing something about it.

The failure by and large to get at all serious yet about (the right kind of) adaptation — a failure this book aims to overcome[2] — is above all a failure *of imagination*. Mitigation (plus limited, merely defensive, shallow adaptation) is easier for our system to contemplate, because if you don't get serious about adaptation you don't have to imagine real change in the world, at least, not within the kind of time-horizons within which our society largely operates. And you don't have to imagine change that actually affects the *nature* of your life. You can pretend that we can simply change what powers our energy-system and enjoy a serene transition... If this was ever possible, it certainly isn't now, as 2022's IPCC reports and 2023's insane off-the-charts weather and temperature-readings have made starkly clear.

This short book, full of punchy, accessible short chapters, returns to reality and prepares the reader for the coming very bumpy ride. This book is for everyone who is no longer willing or able to pretend.

TrAd can feel thus threatening to some: it will certainly shake things up. And TrAd demands attention to the priorities of those on the climate frontlines, those whose lives have already been drastically shaken; especially in the Global South (where more TrAd has thus far happened than in the Global North: because the crisis is felt much more directly and frequently). Adaptation has to date been the poor relation partly because the Global North has been more willing to fund mitigation than adaptation in the Global South. This solves the final element in the mystery outlined above: why to date mitigational action and talk has flourished far more than adaptational. It is precisely because the benefits of mitigation in the Global South for the Global North are more than those of adaptation in the Global South. (Likewise for dissolving the mystery of why the poor in the Global North have been bombarded with information about mitigation, but not about adaptation; because the rich care little about how the poor will adapt.)

What a disgrace. That disgrace too is part of what we aim to mitigate, and start to overcome, in these slender pages. (If, like us, you believe in climate justice now, then you believe in Transformative Adaptation.)

And who is this 'we'? A community of a sort gradually came together, as XR started to get bogged down, beginning in late 2019 and early 2020. People in- and out-side of XR, who wanted to do something more positive, and were attuned to the fact that XR's demands were not being met. Some wise souls among these people decided that there was a need to try to promote and essay Transformative Adaptation in a manner that built on talks I'd been giving since Green House thinktank's pathfinding book appeared earlier that year, *Facing up to Climate Reality: Honesty, disaster, hope*, but in a manner that would be more embodied, more practised and practical, more collective.

That community has gradually emerged as the TrAd Collective. We exist at **https://transformative-adaptation.com**, and in a set of WhatsApp groups. We exist at periodic small gatherings in west England, east England and as the TrAd Village at the annual Green Gathering on the Wales/England border. We exist as a shared and slowly spreading sensibility. Perfect, we ain't. There has been much travail along the way, and still is. But we are trying to make something available, and in a small way (and then maybe a bigger way) to live it.

This book has been made possible and imaginable by the examples and co-operation of numerous TrAd-ders. It is dedicated to everyone trying — against the odds, but nevertheless with history and right and most crucially truth on our side — to make Transformative Adaptation a reality.

By grace and graft, may we together find a way through what is upon us, as wisely as we can. May we grace the future with more sense than humanity has yet managed to add up to.

Foreword

Chris Smaje

When Extinction Rebellion hit the streets in 2018, I was sceptical and slow to get involved. I'd long before abandoned the notion that large-scale centralised nation-states and their governments had the ability to deal with the poly-crisis. Presenting the government with a list of demands seemed futile.

I realised I was wrong when I saw how much the movement brought climate change and other aspects of the crisis to the centre of citizens' attention. Joining the protests myself, I was energised by the ability of grassroots activism to claim space – however modestly and temporarily – from the powers that be. The chant of "Whose streets? Our streets!" has stayed with me.

But for all that, I don't think the reasoning behind my original scepticism was entirely wrong. Dramatising the issues through protest is important, but it isn't enough – and I still think governments lack the ability to deal with the crisis. So it's important to keep moving, thinking, acting and innovating grassroots responses all over the place to the multiple dimensions of the present crisis.

That's why I'm so delighted that this book and the nascent movement that generated it has arisen. To me, they're the next piece in the jigsaw that the climate protest movement started work on. I won't dwell for too long on why. My job here is to get out of the way quickly and let the book speak for itself. But I just want to highlight a few themes that resonate with me.

We need Transformative Adaptation because a smooth, tech-led 'transition' to a high-energy, low-carbon iteration of the modern global economy is no longer on the cards, if it ever was. 'Adaptation' needn't be a surrender to the forces of big energy and big capital. In these pages, it emerges as a radical challenge to them.

Access to land and its capacities to feed, shelter and sustain us rightly looms large in what follows. This is going to be a defining political battle in many places in the years to come. It needs to be done collectively, but in thoughtful and nuanced ways that preserve diversity and build alliances whenever possible. It also needs to be defiant when necessary.

In transforming and adapting to new circumstances, there's a mountain of things we can learn from people in the past who figured out low-impact ways to live and to be. To do so, we need to get over ourselves and our modern emphasis on 'progress'. The point isn't to hark back to a better imagined past but – exactly as the book says – to go 'back to the future', crafting a new synthesis of past practices that shouldn't have been abandoned and present ones that ought to be retained.

Finally, we need to approach present problems not like a simple repair to a broken machine that can be left to get on with the job once the malfunctioning parts are swapped out, but like the complex people we all are, enmeshed in ever-challenging and changing relationships that require our constant attention, love and ingenuity. Something I particularly like about this book is that it understands this and doesn't try to give 'the answers' in the manner of a blueprint or repair manual. The answers – only ever partial and bounded to given circumstances – lie within all of us to seek and enact. Ultimately, they must be cultural and spiritual answers much bigger than any machine, or any protest movement – but they'll be braided out of endless smaller responses to particular and local issues. I find this book enormously inspiring as a way to rise to those challenges, small and large. Whose world? Our world – to sustain us and be sustained by us, to nurture, protect and hand on.

Chris Smaje is a writer, social scientist and small-scale farmer. He is the author of *A Small Farm Future* and *Saying NO to a Farm-Free Future*. He blogs at **www.chrissmaje.com**

Before formally introducing Transformative Adaptation, we felt it necessary to summarise the criticality of the moment to emphasise why both adaptation and transformation are necessary. The science and policy presented is accurate to the best of our knowledge at the time of going to print.

Chapter 1

A Moment of Criticality

Rupert Read

You *know*. Just by virtue of your having made the decision to start reading this book, I know that you basically *already know*. You know that we are in very serious trouble. Too serious for 'Steep emissions reductions now!' alone to be, any more, a mantra that can be counted as remotely serious.

Non-public focus groups I've witnessed in recent years, among people who are not 'progressive activists', suggest that actually many many of those who are not as yet *acting* that much nevertheless *know*, too.

Opinion polls suggest it also. There is a majority who know climate to be a mortal threat, that our governments have not yet got even remotely serious about. Our task, our great work, is to turn that climate majority into a majority who get active.[3]

But the purport of this short chapter is to dwell a little, before we turn to that task, on what exactly the threat is. Just how unprecedentedly serious it is.

All I will do here is very succinctly describe its outlines, and offer you some key references to support this.

The most obvious contour of the threat is the climate more-than-emergency.[4] To understand the nature of that mortal threat, there are just so many sources I could mention. If I had to pick one, it might be Chapter 1 of the book *Deep Adaptation*,[5] entitled, 'What climate science can and cannot tell us about our predicament'. I'm pretty proud of this piece of work, but part of the credit, and of its authority, rests neither with Jem Bendell nor with myself (its authors), but with a kindred spirit, a climate scientist who thoroughly reviewed – and in fact helped research – the chapter. This person was unwilling to put their name to it, because they were worried about the potential blowback from colleagues from the starkness of the truth delivered in that chapter; even now, scientific reticence is holding us back, collectively. They strongly encouraged us to go ahead and publish it.

The broad outlines of the climate crisis are now known to those who are even slightly paying attention. In particular, it is little more than a cruel joke now to hold out to humanity the prospect of remaining below the 1.5°C semi-safe upper limit to global overheat, that governments promised us. That ship has sailed.[6]

But it is important to understand that climate breakdown is *only a symptom* of a deeper and wider crisis: the ecological crisis. For this, the greatest single source is the Stockholm Resilience Centre's 'Planetary Boundaries' framework. In this framework, climate is only one of nine planetary boundaries. Most of the nine boundaries have now been crossed. This, and its emotional implications, are explored beautifully in David Attenborough's Netflix documentary, *Breaking Boundaries: The science of our planet*, though since that show was produced, more boundaries have been documented as crossed.[7]

The ecological crisis is in turn a symptom of what in an important sense is a yet deeper crisis. This is a crisis of political economy, of democracy, and ultimately of our civilisation itself.[8]

The crisis of political economy is essentially that we live in a system which mandates endless economic growth (thus necessarily crossing more and more ecological boundaries), without even satisfying people's most fundamental needs (which, roughly, are for nature, connection, meaning, security, and community). This system drives levels of inequality that would have made Roman emperors blush; for while many of us now live levels of consumption that in previous epochs were reserved only for a minute elite, more of us still subsist at levels of extreme poverty that would have been recognisable across history – while the new elite have access to wealth that utterly dwarfs, by many orders of magnitude, what ancient elites could draw upon.[9]

As the very habitability of our Earth teeters, unimaginable 'profits' continue to be made, and capital attempts to commodify life itself (e.g. forest ecosystems; on which, see the important work of Biofuelwatch, as well as my own academic efforts).[10]

The crisis of democracy can be summed up in this way, riffing on Gandhi: democracy would be a good idea. The actually existing 'democracies' are in obvious respects massively preferable to the tyrannies crowding our world; but they are not spaces where the people rule (which is the etymological meaning of 'democracy'). They are in effect largely part-corrupted oligarchies with a veneer of electoralist legitimacy. Democracy badly needs upgrading. (See on this inter alia Chapter 4 of my book *Parents for a Future*.[11] See also Jon Alexander's important book *Citizens*.)[12]

The overarching crisis of civilisation includes these crises but ultimately they are in turn symptoms of it, of the deepest crisis of all that we are subject to, which I would call existential,[13] philosophical or indeed *spiritual*. (I sought to delineate this crisis in *This Civilisation is Finished*.) Our crisis is ultimately one of values, of the meaning of life itself. We are sacrificing the future on the altar of 'prosperity', without actually managing to produce what is of value, without managing to actually *prosper*; this incredible self-defeat reeks of emptiness. We have lost a grip on what the economy, and even existence itself, is for. But this can be recovered, and more and more of us are — sometimes blunderingly, but good-heartedly and with determination — seeking and recovering it. Disclosing it. We can find great joy in the very struggle that is now upon us, in the present moment made more precious in the revelation of its very fragility, in each other as we recognise common purpose and travail, in the glory of life. If I can't dance, it's not my spirituality. Any sane response to our predicament will be saturated by grief, but the grief in turn will be founded in love – that can re-express in joy.[14]

And above all perhaps, we can find great *meaning* in all this. The levels of nihilism, anomie, purposelessness with which our failing civilisation is saturated; all these are intensely vulnerable to finding our purpose in the common struggle to prevent our growing, shared... vulnerability from sweeping us away. And in the darkness of this time, that is a deeply encouraging thought.[15]

As we turn to face how vast the much-more-than-problem that we face is, how the devastating world-ending climate more-than-emergency is only the tip of the series of icebergs I have just run through, it is obvious that the 'game' now concerns a meaningful, massive endeavour to adapt to what is upon us in a way that takes its dimensions seriously, and does not pretend that we are dealing with anything that can be 'solved'.[16]

We must (therefore) henceforth take adaptation *at least as seriously* as mitigation/prevention.

Chapter 2

The Age of Adaptation

Morgan Phillips and Rupert Read

Our world is climate damaged. Irreversible harms have been inflicted. More damage and more harms lie ahead. The world's mitigative efforts are limiting and will limit the impacts, but – at the time of writing – they are still way too little, way too late. They will not secure us a future of low impact climate change. What lies ahead is slow violence and sudden shock. Neither will be spread equally, but the impact on the planet will be, in aggregate, severe.

What is in play is breakdown, not mere anodyne vague 'change'.[17] And this means that all humans and non-humans will feel the effects of the unnatural global overheating and climate chaos now unleashed and be forced into a mode of adaptation. Adaptations will be spontaneous, in reaction to changes being felt in real time. But they will also be planned, in preparation for the changes underway and to come. There is no guarantee, however, that these adaptations – whether spontaneous or planned – will be effective, or even benign. Many adaptations will turn out to be insufficient, and many will be *mal*adaptive causing collateral damage locally, regionally, and globally. The 'how' of adaptation matters.

After years of intransigence, the Western environmental movement has begun to embrace the need for adaptation. Now, when people contemplate human-caused climate change, they usually no longer adopt a purely mitigation mindset, where their only response is to shout 'mitigate, mitigate, mitigate'. An adaptation mindset is slowly being adopted as people recognise that their response now has to be mitigate *and* adapt.

As mindsets shift, and interest grows, new framings of adaptation are emerging. Into this space comes 'Transformative Adaptation' (TrAd). It is a nascent philosophy, movement, and practice that is fit for the *age of adaptation* that the world has now entered.

TrAd recognises that dangerous climate change is here, unjust, advancing, and already beyond safe limits. In acknowledging this reality, TrAd accepts that while efforts to limit climate change are essential, ongoing, and imperative, the impacts are large enough now to touch every human being (and non-human being) on the planet. We cannot escape it, and so... *we are all climate adaptors now*.

In this book the TrAd movement puts forward the case for being *transformative* as we adapt.

We explore the how and the why and build towards a declaration that serves as a launchpad for what we hope will be more thinking, writing and (most important) doing in the years to come.

TrAd holds open the possibility that human-triggered climate change can – through societal transformation – be halted, and that it will not go 'runaway'. TrAd holds open the possibility that ecological breakdown can – through societal transformation – be reversed; and that the complexities, wonders, and diversity of the natural world can be regenerated, restored. TrAd holds open the possibility that the extremes of economic inequality and injustice can – through societal transformation – be corrected. TrAd therefore holds open the possibility that societal collapse can – through societal transformation – be avoided, even yet, and that vibrant successor civilisations can emerge to replace today's destructive and hegemonic model of Western civilisation.

TrAd is underway in the Global South and North; it builds on advances in permaculture, agroecology, eco social anarchism, the Transition Towns movement, Ecological Swaraj, NVDA (Non-violent Direct Action), and co-liberation. This book will make the case for TrAd and bring it to life through case studies of practical, replicable TrAd initiatives around the world. It will visit Climate Emergency Centres, community farms, land-based protest movements, and citizens assemblies. The writing will be accessible and direct, passionate, and – we hope – transformative.

We have tried to achieve a balance in this book. We want (on the one hand) to give the reader a direct sense of the raw energy, crackling sometimes – like dispatches from the frontlines of intellectual and practical struggle – that has characterised the emergence of TrAd *as* it emerged, and (on the other) offer the reader a sense of what TrAd has become and is now.

We've tried as editors to balance preserving the pieces in this book that have been reprinted as historical documents and making them fully contemporary to the time of publication. We've tried to exemplify the spirit of adapting to the moment, of transforming as we go, but without doing violence to our roots... As a result of not wanting to eliminate all sense of TrAd as something with a history, the short chapters of this book, written by different people, are sometimes in certain respects slightly repetitious.

While editorially removing some repetition, we have kept it when it is essential not just sometimes to drive a point home, but to give the reader a sense of the character of the moment and context of composition of the chapter in question, and of what its author was trying to achieve by it. We hope the reader will forgive this, in a good cause; but, obviously, you should feel free to skip bits when you are confident you have already got the point, or indeed when other portions of the book hold more vivid interest for you.

Let your experience of this book be pragmatic, and pull out of the book what you need from it. *Be* transformative and nimble in your reading of it...

What follows is an unfolding of TrAd in a series of semi-independent chapters and – to set us forth on the chapters still to be written – the TrAd Declaration, a work-in-progress (it is always transforming, adapting to the changing times and our changing understanding!) from the TrAd Collective that seeks boldly and invitationally to set the tone for the path that together we will make. For pathwalkers today know that there is no path, in the literal terra incognita of the new emerging Earth. *We have to* make *the path as we walk together.*[18]

The 'TrAd' conceptualisation of Transformative Adaptation was first described by Rupert in the opening article of a special series commissioned by Permaculture *magazine in 2021. The chapter that follows is a re-production of that original piece (with one new section added near the end, and some small emendations by way of updating throughout), published on 31st January 2021, in Permaculture #107. Rupert presents TrAd as a new approach to the climate crisis, one that unites practical mitigation with a kind of targeted protest or protection.*

Chapter 3

Transformative Adaptation

Rupert Read

Extinction Rebellion (XR) has done fantastic work moving the dial on the ecological emergency. Meanwhile, the Transition Towns movement, regenerative agriculture and permaculture have long been working on the ground to seek the changes we need, bottom-up. Is there a way to bring these two approaches together?

XR's approach to the long emergency, like virtually everyone else's in the 'environmental' movement (including strikingly Just Stop Oil's), has focussed to date on so-called 'mitigation' — on reducing climate/eco damage (ideally, to zero) by reducing emissions and habitat-destruction, by putting pressure on the government to act.

This is quite simply no longer tenable as the sole or even main objective. Too little time is left; and government (and the system) is too profoundly resistant to doing the right thing. Trying to fill the jail cells (with a diminishing number of activists who are willing to do so) isn't enough; and 'mitigation' in its technical sense just is not a sufficiently encompassing objective.

Many of us have been aware of this for some time — and this awareness can no longer be kept at bay. The 'Theory of Change' the radical flank has been using is not enough.

We need to embody the transformation that we aim to bring. We need to be fully addressing the crisis, which has gone too far to be addressed through 'top down' measures alone.

It is far too late for 'mitigation' alone; we need to make adaptation, often in the form of bottom-up action, central, too.

Adaptation comes in three forms:

Shallow Adaptation, e.g. building higher sea walls. This involves no significant psychological change, just business as usual while trying to cope with a deteriorating world. (The Intergovernmental Panel on Climate Change (IPCC) calls this 'incremental adaptation'.)

Transformative Adaptation (aka 'TrAd'), e.g. restoring wetlands/ mangroves, using appropriate technology, living closer to the land. Transformative Adaptation is transformative because it is the route to system change. It requires willingness to undertake major psychological adjustments away from what has been 'normal'. The IPCC calls this 'transformational adaptation', though what is meant by this phrase is often more modest than what we have in mind as TrAd. Part of the motivation for organising through the lens of Transformative Adaptation is to turn this inspiring phrase into a reality worthy of it.

Deep Adaptation, e.g. moving coastal cities inland and reducing their scale. Deep Adaptation is adaptation to collapse. To a future where our existing society is going to be swept away. It requires massive psychological adjustments.

I am strongly in favour of Deep Adaptation, which I regard as an essential hedge, a crucial precaution, and so I have co-edited a book with Jem Bendell on it. But on its own it is not enough. It might even run the risk of being a self-fulfilling counsel of doom.

The Three Forms of Adaptation Compared

Shallow/merely incremental adaptation is worse than useless on its own, because it pretends we can keep this civilisation stumbling on as is without real change. We can't. The longer we try to do so, the further we go off the cliff.

Deep Adaptation (DA) is necessary: but I hold that it should be viewed as an insurance policy against worst-case scenarios, not as the whole goal/programme. In particular, it risks debilitatingly assuming, with a knowingness that we cannot (yet!) know is justified, that collapse is definitely coming. DA works best if conceived rather as in alliance with TrAd. DA is then taking precautions against a possible/likely (not certain) collapse.

Transformative Adaptation (TrAd), as reconceived and emboldened here (building on the origins of the concept in academia, the UN, and the work of the Green House think tank – see our edited book, *Facing up to Climate Reality*),[19] breaks new ground in being potentially willing to use NVDA (Non-violent Direct Action) protectively as a tool in pursuit of these aims where appropriate, one (key) tool among others.

Both Transformative and Deep Adaptation make it essential that we confront what I call 'the great sorrow'. It is too late to accomplish what my teacher, Joanna Macy, calls The Great Turning,[20] even on an 'emergency' basis, in a way that will prevent great suffering. Great suffering is coming. Actually, it has obviously started and far more is coming. This is certain because of the onward momentum of the system we have and the bad feedbacks to some extent already baked into the planetary system.

A Vision For This

Civilisational decline – the ending of *this* civilisation – is inevitable. That future is already here only it's not evenly distributed, yet ... In other words, some parts of the Global South are already experiencing collapse. The extremely hard task is to help each other morph gracefully into a better civilisation as we undergo energy descent. That process (which is very likely at some point to proceed via civilisational collapse, but doesn't yet have to) will primarily be a process of adaptive relocalisation; a process that might have been to some extent jump-started by Covid. This is Transformative Adaptation. TrAd is a win-win-win: we mitigate the effects of dangerous climate change, we work with Nature not against her, and we transform society in the direction it needs to transform anyway. TrAd enables us to cope with the deterioration that is already baked in while potentially improving our society and our future.

If we pull this off without collapsing altogether, then we get to keep the best of what we have (including global communications interconnectivity) while rebuilding community and insuring ourselves against over-dependence on long, uncertain, polluting supply lines. With TrAd-centred 'glocalisation', our world can be the best of both worlds!

TrAd goes 'back to the future' – it returns us to aspects of traditional life that should never have been abandoned, while keeping those features of life today that ought not be lost and that are completely compatible with One Planet Living.[21]

If necessary, we can (and will) do this for ourselves; we have to try.

How at a Level of Fundamentals is TrAd Different from the Transition Towns Movement?

It isn't. They are natural allies. But this needs to be both realised and real-ised. To achieve (and defend) TrAd will involve politics in the broad sense of that word; it will probably at times to come potentially require serious non-violent direct action, e.g. guerrilla gardening at scale. Think landscape-scale agriwilding, not necessarily with consent from 'landowners'.

Or defending the taking back of land for the landless... These ideas, and starting to implement them, are something that the experience of XR and the radical flank more generally (including learning from its blunders) can bring to the likes of the Transition Movement.

TrAd actions are likely to be most effective if they are the change we want to see in the world. If they are beautiful and powerful, if they make sense to people, if they try to create something and not just to hold something back. The most obvious ways therefore of narrating TrAd will involve the growing of food or activities similarly fundamental (to our lives) and the birth of a vision of a good and liveable future, despite everything. Ergo, TrAd combines the practical goals of permaculture with the methods of XR. To embody senseful acts of beauty.

A Way Forward: A Rebel Alliance?

We need an alliance between the likes of XR and the likes of the Transition Towns movement. If someone wants simply to go and grow food and create a community etc., great – but lots of people were of course doing that before XR. And what Transition folk have found again and again is that the powers-that-be stymie change at the point where something genuinely radical is being considered. It's time to stop letting such 'No' be taken for an answer.

We have an opportunity here to unite many of the hitherto-separate movements seeking to model and create a healthy planet. That is the golden thread: that is why we seek out actions that embody the story of system-change via an alternative system that we model AND that challenge the existing system at the same time. A liveable future is going to require relocalised living AND, once in a while, civil disobedience. What will be truly transformative is if these get done simultaneously.

Three Pillars of TrAd

Land

By this we mean the land, the water and the air. And we mean place: land is not just a transferable commodity; land is typically unique. What the land IS depends partly on where it is, and on its history.

And much land is or should be a commons. It is not separate from the communities that depend upon it and that guard it.

People need land. Many people want to be closer to the land, including working it. Part of Transformative Adaptation is without doubt going to be returning land to people.

Community

Fundamental to TrAd is an assertion of the importance of real community – which means, most importantly, actual geographic community, collectivity of people in a place – over the pseudo-individualism that dominates the popular image of our culture and is killing us (as can be witnessed very clearly in relation to Covid, but is equally true in relation to climate). Community requires connection; with each other, and also with the Earth and its many other beings. It requires co-liberation: our freeing ourselves, together (on which, see the next chapter). Ultimately, it requires common, upgraded democracy: this is a crucial part of what we want to achieve (and where feasible, it is also a great way to achieve it).

Transformation

The TrAd vision values tradition. It values age-old wisdom, and does not fetishise 'progress'. But the vision is clear-seeing that we cannot now survive, let alone flourish, without transformation. Transformation is coming; our current system/civilisation is finished and attempts to prolong it will only worsen a crash (thus bringing about a brutal transformation). The transformation that is needed is one that is intelligent, deliberate, recognising this truth about the best-before date on our current way of living. Such transformation will be about going back to the land and moving forward into a new future (a future where we will use less energy, but probably by far most of the energy we use will be genuinely renewable). The existing system is unlikely to facilitate such transformation. Thus, while we should aim to get transformation happening through governments, the UN, etc., we should also be realistic enough to recognise that most of it will probably have to be led 'bottom-up'. And may well, as intimated above, require civil disobedience to be made possible, sometimes at scale.

In summary: We can only expect people to care about our shared global environment BY taking care of their fundamental needs – for meaning, for connection, for food and drink... Ergo the need for transformation, for community, and for land! Together, the fundaments of TrAd, which one might picture as three interlinked spiralling pillars of a tripod...

Conclusions

If we only talk/do Deep Adaptation, we'll be accused of giving up. A gulf will open up between the localisers/'crusties' and political/activist types. But can't great, unprecedented strength be found if we were to really bring these two distinct types together? TrAd is a way to do that. It's never been done before at scale. Now is the time.

We should drop much of what we think we know, including the rash assumption that we KNOW what the future will hold. A humble, precautious, unknowing attitude to the future is exactly what TrAd cultivates. We need to move into an uncertain future in a way that does not presume that we know that the future will be good ... or bad. TrAd protects against the worst – while inviting and seeking to manifest the best. It could enable us to transcend the 'environmental' movement – and start to build an eco-logical future.

Rupert was joined by Skeena Rathor for the second article in Permaculture *magazine's 'TrAd' series. Amongst many other things Skeena co-founded the Guardianship and Visioning team of Extinction Rebellion; she's also known for her work as a therapist and mentor. The chapter that follows is a re-production of Rupert and Skeena's Paulo Freire inspired piece that appeared in Permaculture #108, published on 30th April 2021. This chapter outlines how the very process of transformative adaptation to climate breakdown can also enable us to reclaim our interdependence and step into our full power as a species.*

Chapter 4

How We Will Free Ourselves – Together

Skeena Rathor and Rupert Read

This then is the great humanistic and historical task of the oppressed – to liberate themselves and their oppressor as well. The oppressors who oppress, exploit and rape by virtue of their power, cannot find in this power the strength to liberate either the oppressed or themselves. Only power that springs from the weakness of the oppressed will be sufficiently strong to free both.

Paulo Friere

Readers of this book are well aware that we face a vast ecological challenge. That challenge has to be responded to socially and (at least with a small p) politically. For there is no way we get through what is coming without, at minimum, a serious rebuilding of community.

In the last chapter Rupert outlined Transformative Adaptation (TrAd) as a viable conception of how to respond to this challenge. TrAd has, in common with the permaculture and the Transition movements, a clear recognition that efforts to 'mitigate' the long climate emergency are plainly no longer enough. Climate disasters are here, and they are going to worsen for a long time to come. The climate and ecological 'emergency' is completely different from emergencies as we have previously known them. It forms a condition that defines our lives. It is not going away. In this marathon, we have to create ways of organising and living together that are viable for the long term.

The funding (if at all) and practising of adaptation to date, has mainly been 'shallow' and incremental, such as building higher and thicker sea walls to defend against bigger storms and rising sea-levels. This is not viable in the long-term, and actually adds to the deadly carbon-excess problem in the process. As the previous chapter laid out, we have to adapt instead to a climate-chaotic world in ways that help to mitigate/

drawdown, that work with Nature rather than against it, and that transform our communities and society in the kinds of ways that it needs to transform.

This is Transformative Adaptation. And it will inevitably involve transformational socio-political change of a quite specific kind... co-liberation.

In the rest of this chapter, we succinctly undertake the following two (related) tasks:

Offering some examples of what TrAd could be or is;

Explicating how, in order for TrAd to be realised, we are going to have to real-ise a social and political ecology.

That is to say, we are going to have to find ways of embodying and developing the literal vital insight in the quotation above from the great pedagogue, Paulo Freire. We are going to have to get serious about overcoming the selfish fantasy that we can achieve freedom and self-realisation at the expense of others.

Skeena Rathor has dubbed the programme that develops this realisation 'co-liberation'. Co-liberation calls for integrating self, community and institutions toward greater freedom and greater belonging. Co-liberation requires liberation work connected to power and leadership. If your freedom is integrally related to mine and mine to yours, so is my power, my thriving and my leadership.

Thus, how we co-liberate is informed by and informs how we do shared leadership and shared power. All liberation is, properly speaking, co-liberation, where the 'co' is a matter of reclaiming our interdependence as a positive thing.

Co-liberation requires recognising that all liberation is interdependent — that liberation depends on growing our common shared safety, freedom and thriving. Working with the scientific and spiritual truth of interdependence, this pathway invites a shift in consciousness.

The need to free the oppressor and oppressed at the same time requires us to acknowledge that we can all be both in changing contexts and circumstances. Co-liberation seeks to free the parts of us trapped in repeated oppressor-oppressed behaviours at an individual/group and systemic level by creating 'power-with and power-up' practices.

The practical philosopher Hannah Arendt has described power as the ability to not just act, but to act in concert. TrAd, as mediated through the novel conception of co-liberation that we are helping to develop, can enable us to figure and realise such power. Power thus reconceived is not something to shy away from. It is a wonderful thing. As activists and activators this asks us to become ready to act in power and take power rather than resist power.

There won't be anything worth calling 'transformation' unless there is an effort to co-liberate, in intentional communities and in organisations and movements, integrated where possible with the broader communities and society around them. And, let's be honest, there won't be Transformative Adaptation at scale unless we are sometimes unwilling to take no for an answer. In other words: what TrAd adds to standard permaculture includes an upfront willingness to do things which are not strictly legal or in obedience of the current system... We draw on the experience of Extinction Rebellion, and we hark back to great examples from the past such as the Diggers or the Kinder Scout mass trespass, and from the present such as the epochal Indian farmers' movement (that in 2021 succeeded, remarkably, in drawing big concessions from the Modi government) and the Brazilian land-for-the-landless movement. If we are going to succeed at transforming in time to beat collapse, then we are sometimes going to need to do the kind of thing they did and do.

We are going to need to press the hand of local authorities or businesses or NGOs; we are going to need to do the kind of thing that Transition Heathrow (aka Grow Heathrow, growing food on the site of a would-be runway) did, at scale; and do guerrilla gardening at scale, and pop-up allotments (that we defend) at scale...

We need to step into our full power in such ways.

When we do these things non-violently, evoking the spirit of Gandhi and King, and evoking in those who might seek to stop or arrest us, the difficulty of an oppressor-role that they don't really want to be in, we are finding a way of creating the dance that can evoke co-liberation.

We skip forward here to Permaculture #111, published on 31st January 2022. Morgan highlights two examples of how TrAd might manifest in the real world. The first draws on his time as Co-Director of The Glacier Trust and the work they do to enable adaptation in rural Nepal. The second visits Walworth Garden in south London and promotes the idea that Garden Centres might just emerge as localised hubs of Transformative Adaptation.

Chapter 5

Agroforestry Hubs

Local Models of Transformative Adaptation

Morgan Phillips

The dramatic upsurge in climate protest that has happened over the last five years is having an effect. The commitments governments are making to climate action have strengthened. According to UN Climate Change, we can now expect global average temperatures to increase to 2.7°C above pre-industrial levels this century. It would still be a catastrophic level of warming – and it is still very much an 'on paper' prediction – but it would be a full 0.5°C better than the 3.2°C the UN was predicting in 2019.

Protest works, it will go on working, so it should go on happening. All power to Extinction Rebellion, Fridays for Future and all the backstage actors who have been agitating away in the darkened hallways of power. That the world will only overheat by 2.7°C is, of course, no cause for celebration. It would be nothing short of disastrous. The planet is already 1.2°C warmer than preindustrial levels and the impacts are horrendous. Thanking world leaders for keeping temperature increases to 2.7°C rather than 3.2°C, is like thanking them for bringing you a piece of toast cooked at setting nine instead of setting ten: burnt toast is burnt toast.

The continued failure to come up with a plan that gets anywhere close to keeping temperature rises below 1.5°C makes one thing very clear: no amount of wishful thinking about 'green growth', 'carbon dioxide removal', and 'decoupling' is going to make the endless pursuit of economic growth compatible with a stable climate. Government scientists, economists, and spin doctors have been trying to find a way to square that circle for decades and they keep coming up short.

And yet politicians and media commentators continue to attempt to reassure us that they are not going to burn the toast. They tell us that they have the situation under control, and that economic growth will, in the end, solve the climate and ecological crisis. They are lying to themselves and to us, but they are losing credibility rapidly. This is the decade where the logic of pursuing endless GDP growth – and the consumerism it relies on – will fall apart. Neither delivers on its promise of a good life for all, and both are doing more harm than good. This explains why ideas like degrowth, doughnut economics, post-consumerism, wellbeing economics, and now TrAd are starting to capture the imagination.

These ideas will continue to develop over the coming years and will attract new and bigger audiences. They provide us with the alternative visions of progress that the world is calling out for. Protesters and activists must remain hopeful and keep up the pressure; we all should. We need to amplify ideas that reach for genuine change, ideas that tackle root causes as well as symptoms, and ideas that co-liberate the oppressed, and the oppressor. The clarion call 'end fossil fuels' needs to be flanked by calls to end more entrenched ideas like consumerism, infinite economic growth, and potentially even that most dubious of constructs: 'Western' – so-called – civilisation.

This book brings the emerging theory and practice of TrAd to the surface. The thinking behind it points the way to a post-consumerist, post-Western, post-growth world. TrAd is a response to dangerous climate change, but it isn't a narrow within-system response. In that way it is distinctly different from conventional, incremental forms of adaptation that – by accident or design – mask and therefore deny the realities of human-made climate change. TrAd is therefore not interested in scaling up the use of snow making machines to keep Alpine ski resorts viable, nor the 'air condition everything' strategies of heat stressed Gulf states.

TrAd and 'TrAdders' are alive to what climate decline is telling us: Western civilisation is breaking down, climate stability and capitalism – in any form – are not compatible. TrAd therefore aligns with efforts to shift paradigms in economics, wellbeing, land ownership, agriculture, education and leisure. In fact, it aims to contribute to these efforts by identifying, imagining and creating 'successor civilisations' that are not only adapting to climate change, but also doing something far more fundamental: steering society away from a consumerist, hegemonic, and – it turns out – fragile model of civilisation that has dominated for so long. This isn't within-system tinkering, it is system transformation.

TrAd Examples

There is a lot to learn from what is happening in Rojava: I am referring to the Make Rojava Green Again Movement in northern Syria (or southern Kurdistan, depending on your point of view), as an emerging example of what we mean by TrAd.[22] The successor civilisation that is establishing itself there is an act of defiance against the Syrian government, ISIS, and the forces of Western civilisation. It gets its strength from the compassionate and cooperative values it nurtures and brings to the surface. It is explicitly seeking ecological harmony, gender equality, resilience, and democratic autonomy.

Rojava is inspiring, but it isn't a utopia, nor to everyone's taste. It is also under huge pressure in what is still a war zone; it may not survive. But if it does, and does so *in extremis*, it will have proved itself, giving those who admire it the confidence to copy and replicate it elsewhere. I have written about Rojava in more depth in two recent books *Great Adaptations*,[23] and *Climate Adaptation*,[24] where I also cover the community-led Agro Forestry Resource Centre movement that is establishing itself in Nepal. This seems to be another example of TrAd and there are elements of it that could be replicated in the UK, and elsewhere in the Global North.

Agroforestry Education

An Agro Forestry Resource Centre (AFRC) is a plant nursery, and a space for innovation in agriculture, community, and climate change adaptation. Community-led AFRCs are run as independent charities, making them not-for-profit, and are governed by a gender balanced board of local volunteers. They comprise a physical building – with a kitchen, accommodation, office, bathrooms and a classroom – that is surrounded by demonstration plots, water harvesting ponds, greenhouses, trees and a kitchen garden. These indoor and outdoor classrooms serve the centre's main purpose: education. Expert staff use them to host regular training sessions for local farmers on various agroforestry and agroecology techniques and methods.

Farmers come from near and far, with some staying overnight to attend two- or three-day workshops that combine theory with practice. Farmers learn how to grow and nurture crops organically, how to intercrop and layer farm, and how to harvest and process high value crops like coffee, almonds, macadamia and hazelnuts. It is worth stating that in these sessions educators aren't simply dishing out facts and direct instructions to obedient learners, they are co-learning.

It is a continuous process of knowledge sharing and discovery; they problem-solve, innovate, and experiment as a group. This is how new adaptation methods emerge and spread.

AFRCs also have outreach programmes that are run through networks of smaller satellite centres. These satellites are dotted around the local community, ideally at different altitudes and spread at different points of the compass. They are managed by lead farmers who are all stalwart members of the central AFRC. This is one of the ways the AFRCs are community led.

The advantage of creating the satellites is two-fold. Firstly, the farmers who run them are happy to host workshops on their land so that staff and AFRC members can come together to learn agroforestry and adaptation strategies that are appropriate at different altitudes and aspects. Regular workshops are therefore held on location with farmers learning from each other and through practical activities. I have attended sessions on bio-intensive tree planting. The attention to detail and preciseness of the methods used are quite something to witness; they are always fine tuning the approach as they learn from previous efforts. Staff often film proceedings too and make videos to share in person and via social media. I have spent many evenings watching various planting or pest management techniques on the phones of farmers I meet in the mountains.

Secondly, the satellites are small-scale plant nurseries where local farmers can come to buy seeds, seedlings, and saplings at a convenient location, saving them what can be a long and tiring journey to the main AFRC hub.

And thirdly, the farmers who host the satellites have a wealth of knowledge and experience. They become informal role models, acting as coaches who pass on advice and wisdom to their less experienced neighbours.

Creative and Collaborative Values

The two AFRCs I have been involved with in Nepal – in Solukhumbu and Kavrepalanchok – have also become central meeting places for farmers who come together to form cooperatives. By organising in this way, farmers have been able to work together to source buyers for the high value crops their training has taught them to grow. The cooperative in Solukhumbu is now 48 members strong, 25 of whom are women. They receive regular training on how to effectively run and govern their cooperatives. They are selling fruit, vegetables, nuts and coffee locally, nationally and, in the case of coffee, internationally.

The creative and collaborative values that underpin the AFRC model and the cooperatives that they host, are in stark contrast to the creeping influence of Neoliberal, consumerist values in Nepal. These compassionate values align strongly with care for the environment at local and global levels, but also social justice causes, which is why the work being done is addressing far more than just climate change adaptation. The prominence and reinforcement of compassionate values is vital to a broader process of social and economic transformation in the communities touched by the work of each Agro Forestry Resource Centre.

Localised Hubs of Transformative Adaptation

Garden centres in the UK and the Western world could learn a lot from the Agro Forestry Resource Centre model. They too could become hubs of TrAd in the coming years. Many garden centres already have their own plant nurseries, and some offer formal classes and informal advice to their customers. But they could be doing more than this. They could recruit lead gardeners in the neighbourhoods they serve in the same way that the AFRCs in Nepal have lead gardeners in the wider community. Lead gardeners could open their gardens up to their neighbours and share learnings on permaculture, organic farming, climate resilience, adaptation and more. They could also grow seedlings and saplings to sell to their neighbours, or through the central garden centre.

The garden centres themselves could also become innovation and training hubs that offer workshops on growing food, climate adaptation, community building, gender equality and social justice. If they have space, they might even offer accommodation for deep dive residential training courses. The only UK example of something like this that I have come across is Walworth Garden in the London borough of Southwark. I spoke to Walworth's chief executive, Oli Haden, and he explained to me the origins and principles of the garden. It is not a garden centre; in fact they deliberately call the 'shop' part of their operation a 'plant centre', to be distinct from a 'garden gnome' version of what a garden is in people's minds, as Oli put it. Walworth is also doing things that impact on the wider community, potentially having a transformative effect on the values, beliefs, and life goals of the people its work reaches. The Walworth team will happily challenge the bastardisation of 'organic' for example, and won't shy away from fighting for social justice locally and nationally.

Oli told me that there are other similar organisations in Pennsylvania, USA, but he wasn't aware of anything else like it in the UK. There may well be, and the potential clearly exists. Why can't every town, city and rural area have its own Walworth Garden? If we could evolve garden centres from the cold, individualistic, transactional places they so often are, to civic spaces where citizens congregate and do the doing of transformative adaptation, they could become hubs for the wider social change that is so urgently needed. Himalayan farmers grow different crops in very different circumstances, but their community-led AFRC model is replicable anywhere in the world. I hope to see something similar emerge in my hometown, it will dovetail beautifully with the multiple mutual aid initiatives that are emerging from the shadows all over the country.

In this book's penultimate extract from Permaculture *magazine's TrAd series we hear from four pioneers of the emerging Climate Emergency Centre/Climate Hub movement that began establishing itself in a locked down UK in 2020. This article appeared in Permaculture #110, in Winter 2021.*

Editors' note: This chapter has been brought right up to date to reflect the changes in Guildford and Talking Tree especially, and also across Climate Hubs more generally, since this piece first appeared.

Chapter 6

Climate Hubs

A Transformative Way

Rachel Bailey, Ruth Allen, Ben McCallan and Suzanne O'Hara

Since July 2018, when Adur District Council in West Sussex declared a climate emergency, well over 300 councils have followed.[25] But many have stalled there, lacking the funds, capacity, imagination, or resources to take practical next steps, and most local authorities are currently off the pace on their climate targets.

The vast majority of UK residents want ambitious action to tackle the seriousness of the crisis we face. More In Common's excellent research for 2024, shows around three times as many people support improved cycling and walking infrastructure – even if it comes at the cost of removing space for cars – as oppose it. Yet you wouldn't know this from the 'war on motorists' narrative that pervades most media, and a small minority who have become skilled at making their voices very loud.

As a result, people generally tend to underestimate how concerned their fellow residents and communities are about climate change, and so become less vocal about their worries, and what they're doing to tackle this crisis. But impacts of the climate crisis are becoming more apparent every year, even at home, and people are impatient to experience change, as well as be involved in driving transformation forwards. With action felt to be too slow by the majority of local authorities, communities are left looking for blueprints for initiatives that they can adapt.

Local community 'hubs' are an effective way to flip this narrative, and done well, normative appeals can be used to show local residents taking action, and thereby normalising the behaviour needed to mitigate and adapt to climate change. The Climate Emergency Centre can provide a template for groups wanting to start their own projects, and the many community spaces inside or outside of this network are working blueprints of any community can take collaborative action. Like the TrAd collective, whose ideas these projects can be seen as embodying, community-led climate spaces are a positive response to the crisis.

These community-led hubs are focused on building solutions, adaptation and resilience to the multiple environmental and social crises created by our current system. The same system underlies the causes of both the social and environmental crises.

With this in mind, environmental and social groups come together so they can work in the same space, cross-pollinate, learn and grow together.

The three principles at the heart of these projects are:

1. A solution focus – for people and planet

2. A local and inclusive focus on meeting community needs

3. A wider network for mutual support and cooperation.

There are different ways to approach the creation of these projects, and it's vital that communities approach it in a way that works for their own community. But there are useful resources to help along the way, such as the *CEC Handbook*,[26] which some spaces who are linked to the network have used, as well as articles such as those by Zero Carbon Guildford, which lays out their experience and advice in getting set up. The best way to get started is to find a group who are operating projects that you would like to recreate, and reach out directly to find out how they got it started, and what pitfalls you should avoid.

Crucial to the roll out and success of community climate spaces is a business model that enables property owners to save up to 100% on business rates while temporarily giving a community space using a 'meanwhile lease'. This allows newly established groups to acquire a free 'home' so they can focus on what they want to achieve, rather than on covering costs. So far some councils have also been hugely supportive and several have offered council-owned buildings for free.

But operating a space doesn't come without downsides, and this includes the short nature of the lease, as well as the responsibilities of running and maintaining a premises. Spaces in Dorking and Preston both closed due to short term leases, as is the case with the Redbridge CEC which operated a large warehouse. ZERO in Guildford has already completed the full term of a two year lease and had a six month involuntary hiatus whilst waiting to open in their new High Street home, which launched in April 2024.

The workload and risks make it crucial that, rather than rushing into securing a building, the priority step for any group looking to build a project like this is to first build a strong community. Creating a strong base and good governance is key, allowing time to understand how everyone involved can work together collaboratively (thereby avoiding future arguments and factions), and an opportunity to secure key skills and create crucial plans – such as volunteer management, building management, and a strong engagement strategy.

The shared vision of these projects is of autonomous community-led centres, focused on meeting local needs and building local resilience, strengthened by the coordination, support and skill/resource sharing provided by a broader web of centres across the country. This vision is held in the knowledge that people hold the answers, within us and between us, of how to evolve not just to survive, but to thrive.

And for all the challenges, many of these projects have themselves thrived, and continue to find new and innovative ways to create security. Bristol now has a huge space, named Sparks, in a former M&S, with neighbours in Bath, whilst old hands Lewes and Seaford on the south coast are building and growing. Projects can be found from Sero Carmarthen in West Wales to Woodbridge Climate Action Centre in East Anglia, Truro and Plymouth in the south west to ZERO Guildford in the commuter belt. There are also hubs that were established whilst not initially realising that similar projects were springing up elsewhere, such as Elmbirdge EcoHub and Cleaner & Greener in Bromley.

Here are a couple of micro case-studies of climate hubs in action.

Talking Tree – Staines

www.talkingtree.org.uk
by Suzanne O'Hara

In early 2020 a disparate group of individuals met through their shared interest in the environment and sustainable living. The group wanted to take positive action to tackle climate change in Spelthorne, focussing on actions to minimise waste, reduce consumption, preserve natural open space and raise awareness and understanding of environmental challenges locally.

The volunteers agreed that having a physical space would be a key priority, and so the idea for Talking Tree was born. With plans to use the venue as a community space for arts and culture alongside focused environmental projects, the team was inspired to name the venue the Talking Tree after the Saxon word for Spelthorne, which meant 'speaking tree' – a place where different groups would meet to discuss important issues. The climate crisis being the issue of the age, it seemed an apt naming choice.

The group registered Talking Tree as a Community Interest Company so all assets are 'locked' and profits generated are used for the benefit of the community. Having drawn up a business plan, the group approached Spelthorne Council in May 2020, requesting help finding an appropriate building. Spelthorne Council were owners of a vacant retail property in the heart of Staines High Street. The premises had been empty for an extended period and the Council saw the benefit in supporting the project to repurpose the building as a positive community space. The team took possession of the keys to the property in early November 2020 and the ensuing months were a flurry of activity with a core group of volunteer labour hammering, plumbing, plastering and painting to turn a dark betting shop into a versatile community venue which opened on 21st June 2021. The entire refurbishment was completed on a shoestring with appliances, furniture and equipment mostly donated or salvaged and most labour being donated free of charge by volunteers who supported the Talking Tree ethos. Chairs and lampshades were recovered using discarded fabric samples and the beautiful fascia was crafted from the wood hoardings which blocked the doorway of the once derelict building that Talking Tree has transformed.

From opening, the venue centred around an on-site café serving a vegetarian menu 'with vegan aspirations', using surplus ingredients whenever possible. Without an on-site kitchen, the core team of batch cooks instead used a local catering kitchen to prepare the soups and main course that were served at the café by our 55+ café volunteers all of whom received training and Level 2 Food Hygiene certification. While Covid delayed the opening of the venue, the team had delivered cooked meals, free of charge, to the community and we were keen to continue a community kitchen aspect to our food offering, once the café opened. The pay-it-forward scheme allowed patrons to buy drinks or meal tokens which could then be used by customers of Talking Tree who, for whatever reason, were unable to pay.

The main exhibition space at Talking Tree has been used to promote the work of local artists, many of whom specialise in the use of reclaimed materials and all of whom represent the environmental ethos of the venue. The art has been available for sale with a portion of profits donated by the artists to fund our work. The stage has hosted regular music performances by both local musicians as well as award winning folk bands such as 'The Wilderness Yet' who included Talking Tree in their tour in 2023. We have also hosted screenings of topical cinema releases and a variety of talks including a memorable presentation by Knepp Estate representatives and a great Q&A with Dr. Gemma Newman, AKA the Plant Power Doctor.

Since opening, the volunteer base has formed a number of working groups and active projects. The Bike Kitchen group refurbishes bicycles for distribution locally to individuals who can use them for commuting and other active transport purposes. The team, of which four members have been funded to gain Cytech bicycle technician certification, also run regular bike repair workshops and has begun to provide outreach sessions at local businesses. Full Circle Fashion hosts a regular pre-loved clothing pop-up and kids clothing exchange on a monthly basis as well as offering a permanent clothing rail in the venue. Funds generated promote the local accessibility to sustainable fashion as well as supporting a waste management social enterprise in Malawi. The Re-Fill Station offers household cleaning products sold by volume and is going from strength to strength. The Community Fridge runs weekly on a Tuesday lunchtime, redistributing an average of 65kg of surplus food per session, sometimes many times more. We host a monthly Sewcial skills sharing workshop, where local people can learn to mend and make, using a suite of Talking Tree sewing machines and the help of some very knowledgeable volunteers. There is an active food growing team also: Talking Tree has two community allotments that offer corporate volunteering days and supports Incredible Edible Spelthorne, a group that was incubated following Talking Tree's Net Zero Innovation Programme – a citizen participation group that brought Spelthorne residents together with local government and academics to plan local climate projects. The group has a further community food-growing site at Staines railway station and has developed comprehensive food growing initiatives at two local schools. Talking Tree's Biodiversity volunteers have developed new planting schemes at a local retirement village and have undertaken biodiversity surveys with recommendations to local businesses about how to improve biodiversity on their land.

Talking Tree has been successful in creating a welcoming venue, projects and programme of activities as well as close working relationships with Spelthorne Borough Council staff and councillors, Royal Holloway University and local businesses such as Bupa, Novuna and Co-op, all of which have helped to fund some of our projects. However, despite our successes, we recognise the need to increase our impact further.

Three years in, we now have a team of 66 active volunteers and a small number of part-time paid staff. Moving forward, our strategy is very much focused on community outreach and education – taking the Talking Tree mission activities out to schools, businesses and residents, beyond the walls of our venue. We are in the process of accessing funds and recruiting new staff to do just that.

Zero Carbon Guildford Climate Hub

www.zerocarbonguildford.org
by Ben McCallan

With time on our hands at the beginning of the UK's first lockdown, a group of us from various local environmental and social groups got together to discuss the problems we were encountering. How could we avoid working in silos and begin to share resources and skills? What was preventing the wider community getting involved in the urgent work needed on the climate and ecological crisis?

Why were so many activist organisations focused exclusively on mitigation, demanding change of a system which has failed and is the primary driving force of ecological collapse? And most importantly, how could we all work together to begin building adaptation and resilience into our community, local economy, and the way we live and work.

We felt that a physical space which could act as a focal point for community organising could be a game changer in engaging people from all demographics. Not only would this allow us to increase education on the climate crisis, whilst providing practical local solutions for people to reduce their emissions, it would also help in creating greater community cohesion. We want informed, deliberative decision-making, centred on the fact that everyone in our borough shares one thing in common – we will all be faced with the same physical challenges, and therefore social impacts, of the climate and ecological crisis.

When we first contributed to this write up we were waiting to complete on our lease. That premises has been and gone, with us completing the full two years of the lease, and now having a new home for ZERO at 168 High Street. We have also won a national award for 'Innovative UK Community Project', won a £350,000 award from TNL's Reaching Communities (note this wasn't from a climate fund, so we're really proud we're gradually helping to mainstream climate action), as well as creating and managing a home energy advice service, which has raised the charity £250,000 in a consortium with the County Council.

We've supported over 40 groups from across the UK who have been to visit ZERO looking for advice and guidance – something we are always happy to support any fledgling group with.

A physical space has undoubtedly helped us be more visible, but the real testament is to the continued hard work of so many dozens of people who keep the space operating, spend huge amounts of their spare time engaging residents, and run projects like our water testing lab, which – in partnership with Water Rangers and River Wey Trust – now not only does chemical testing, but has built to working with the University of Surrey to create a Community Water Lab which tests river water samples for E. coli.

Let's use the skills and strengths we have to build a new system of empowered communities, one that recognises the value of community-led transformative adaptation.

This final piece in the series reprinted (with some expansion and updating) from Permaculture *magazine (issue 112, Summer 2022) concerns how we can get meaningfully involved in world-changing in the wake of governments having failed us.*

Chapter 7

Transformative Adaptation as Part of the Emerging Climate Majority[27]

Rupert Read

This book has been seeking to set out how TrAd fits into the evolving landscape of 'movements'. TrAd can be seen helpfully as how the spirit of XR meets the spirit of the Transition Towns movement, in the wake of it becoming clear that the radical flank is not going to succeed in forcing large-scale political change from the top down. I am going to suggest here that TrAd needs to be understood as a new *moderate flank* to both XR and Deep Adaptation (DA).

TrAd and DA overlap to a great extent. The crucial difference between TrAd and DA is that TrAd positively aims at preventing collapse.

This book has so far sought to encapsulate the TrAd vision, that is waiting to be made on the ground everywhere, at this literally vital moment...

Doing adaptation right can transform our response to the ecological and climate emergency (not to mention our response to Ukraine and to the power of petro-dictators), enabling us to survive and perhaps flourish.

TrAd offers a deep reframe for the long term — for real long-term thinking/acting.

This chapter takes us deeper into that reframe. By investigating the way in which TrAd can helpfully be seen as part of the new post-XR *moderate flank*, or, as we are now more often calling it: the emerging *climate majority*.

The Profound Failure of the COP System

At the start of COP26 in Glasgow, Boris Johnson talked about it being 'a minute to midnight'. Very well. By that reckoning, it must now be past-midnight. For the clock can't always be poised at just before midnight; the can can't be forever kicked down the road without that kicking having consequences.

Sometimes, one has to admit that time has passed, opportunities gone, forever.

The COP at Glasgow was the one that was supposed to fulfil the mandate of Paris, and the COP that happened to be timed to co-enact the post-Covid resetting.

Expectations were high. They turned out to be very unmet.

And so: You can forget limiting global temperature rises to 1.5°C above pre-industrial levels. Climate Action Tracker reckons that Glasgow puts us on track for 2.7°C of global overheat![28] An outcome along those lines will be terminative of our civilisation.

Glasgow is where the 1.5°C target was taken off life-support. It was the moment when 1.5°C passed into the rearview mirror. For the only slim chance there was of achieving staying within 1.5°C was for the Glasgow COP agreement to be a historic triumph. Whereas actually it was a damp squib.

Which means that, to be blunt: a lot more people are going to die, and a lot more wildlife is going to be extinguished.

The endless repetition of 'Keep 1.5°C alive' doesn't make it true. On the contrary. For the truth is that this mantra is just another forlorn version of '1 minute to midnight'. It keeps alive a fake hope, it guarantees burnout, it fails to mobilise on the basis of the terrible truth: that we are moving into climate breakdown, and cannot avoid centring preparedness, in the face of what is coming.

Their 'Yes we can!' is a lie. That can stops here...

It won't do to seek to divert the can straight onto COP27, COP28, and so forth, in an 'endless' stream whose endpoint is obvious: nemesis. Instead, a reckoning must be reached.

The Needful Pivot Towards Adaptation

What are the main implications of the above? There are several; a key one is the need to blunt the coming climate chaos by making ourselves and our communities more resilient, ground up. The cultural significance of the passing of 1.5°C is that it marks the point at which it is no longer possible for billions of us to adapt to climate damage within today's forms of economics and politics. Economics and politics needs to be fundamentally re-imagined because the +1.5°C overheating world is a very different place. As Johan Rockström puts it: it is a planetary boundary.

We now clearly need, as this book has insisted, to effect a big pivot towards adaptation. Doing so can break the spell of our current inaction in face of the emergency, and to *land* our growing vulnerability.[29] For adaptation, if it is really transformative[30] (and deep),[27] has at its centre the psychological transformation that comes from accepting that vulnerability. And that acceptance is in turn only fully possible if we make it real to ourselves both by acting appropriately (building resilience, engaging in adaptation-activities, is the best way to effect that shift) and by being willing to go through the anxiety and grief[31] consequent upon really facing the abyss[32] that is opening up before us. Doing so can make available an energy and a power that has been tapped by Greta and by XR, but whose greatest potential remains in my judgement still very much in our common future. Truly facing up to ecological reality and undergoing the pain that that facing-up inevitably brings will in my view very likely lead to much larger movements[33] of transformation and adaptation than we have yet seen.

We need to learn from the Global South, where, in some cases and places, as Morgan Phillips wrote about in Chapter 5, TrAd is already being practised, in the face of and because of the greater vulnerability already experienced there.

To land *our growing vulnerability*, we need to turn back towards the land.

The *Post*-Extinction-Rebellion [XR] Movements Currently Taking Shape

The key emerging movements[34] in this moment tend to accept government-failure as a fact and seek to make change more directly. They have, I would argue, more game-changing potential[35] for raising climate consciousness and changing the wider world than even XR did. For they are potentially more inclusive, having lower barriers to entry (not expecting arrestability, nor any broader sign-up to a so-called 'progressive' agenda or identity); and because they are more positive in their approach (not demanding of others to fix the crisis, but seeking to address it ourselves, directly).

We have to be aware of the motion of history itself: of the effect of experiencing the April 2019 rebellion together, for instance, as this country did. If we simply repeat the same kinds of moves in 2024-25 as worked in 2019 or as were tried in 2021-23, we miss the adventure, miss the moment that is nigh, and in fact risk frustrating and boring people. We have to offer novel ways forward, including things for people to do collectively that will really change things. I would argue that XR, operating as what is called a 'radical flank' (to the then-existing environmental movement), opened up a space brilliantly in 2019. *The point now is to occupy that space.* By way of an emerging 'moderate' flank that potentially huge numbers of people might join – mostly people who struggled to identify as XR rebels, even after XR's 2023 change of strategy (to stop disrupting the public), but who do want to change the world directly. Operating in particular through the two most crucial massive areas of society that virtually all of us operate in: in our workplaces,[36] and in our geographic communities.

Readers of this book are perhaps most likely to be interested in acting in your community, your neighbourhood (or creating a new community). Permaculture and agroecology are *expressions* of resilience and adaptation. (On this point see, especially, the next chapter on Community Climate Action.)

Here are some key examples of this emerging mass moderate flank, examples picked in part for their being expressions of this:

Wild Card, a smart campaign with Chris Packham on board, seeking to rewild the royals' vast lands, and in doing so to draw attention to the terrible facts of land-ownership and the dreadful reality of property law in this country.[37]

Transformative Adaptation (aka TrAd) itself, which of course seeks to lead thinking on the changes needed to cope with climate/ecological breakdown, and ultimately seeks to help lead transformations of our land/communities.

Climate Hubs, one-stop shops acting as hubs for local people to find strategies to mitigate and adapt, in the face of social and environmental decline. See the previous chapter on Climate Hubs. Why not create one where you are, if plans for one are not already emergent?[38]

Trust The People, which in lieu of government and councils having created sufficient citizens assemblies, seeks to create them on the ground across the country.[39]

And now here's an intriguing fact about these outfits: all have been formed in the first instance by former XR personnel...

W(h)ither XR?

Has XR done its job? There is a place for ongoing work by smart radical flanks seeking to push the envelope of what seems possible. And it is intriguing to note that XR's new strategy in 2023 was itself a big step in the direction of the new moderate flank strategy. But individual rebels need to ask themselves whether their time is better spent in big (but almost certainly not big enough) protests pointed at power or in making real the greatest legacy of XR: the emerging mass, distributed moderate flank that I've described so far in this chapter. For if this new moderate flank, this emerging climate majority, with transformative adaptation at its heart, becomes as huge as I believe it can — if not just thousands but many millions come to be part of it — then that will produce a whole different kind of pressure on government. It will in fact alter our whole political culture.

Moreover, it makes less sense than ever for the likes of XR and JSO to persist in having demands that focus on mitigation (i.e. on greenhouse-gas reduction) but that do not take adaptation seriously. As we've shown in this book, it's simply too late now not to centre adaptation: adaptation of the right kind, that will truly make us stronger in the face of the coming weather chaos, and that is transforming our way of life in the direction it needs to anyway, to help us live lighter and happier on the Earth. The pivot toward adaptation needs to encompass all our movements.

A Place to Land[40]

In this chapter, I've sought to place transformative adaptation in the context of the moment we are in: a long moment of despair about governments' definitive failure to do enough to arrest our descent into eco-driven civilisational meltdown, and a moment where more and more people are seeking agency to act as we can ourselves, about this.

For, if enough of us do this, in ways that are broadly aligned, then change at scale *will* be the result.

Political inaction and posturing at the macro level does not have to prevent us acting meaningfully in the kinds of ways that the 'climate majority' organisations described ARE DOING.

So, if you're looking to up your level of action in this vital decade, as I expect you are, I hope you will consider joining one of them. Or forming your own. Thus operating where you live in whatever way is most suited to your circumstances to real-ise a form of adaptation (to the climate decline that is here and that is baked in) and of transformation (of how we live, and ultimately of who we are).

Enough of us doing this, as we are increasingly doing, will give us collectively a place to land. Places to turn into land that is ours. Places to come down to Earth.

The next chapter, by Joolz Thompson, was written in 2023 for this book. Here, with reference to Community Climate Action, Joolz reflects on a realisation he and many of us are now arriving at – there is a need to take hold of the situation, ourselves, in our communities.

Chapter 8

TrAd and Community Climate Action

Joolz Thompson

Let's start with the snappiest definition I am aware of for the TrAd Vision:

A world where transformational change to ourselves and our systems has allowed us to adapt and thrive in balance with all life, with earth.

One point five degrees Celsius is dead.[41] We're on course for 2°C (or more) and climate-deadly emissions are just continuing to rise. This information has made the news,[42] but there is little action on the reality of our situation.

Our pale blue dot, like the Titanic, is sinking. Spaceship Earth[43] is hurtling towards the sun. Yet our 'crew' are doing little to change course...

A lifetime of inaction on the climate and ecological emergency means our communities must now adapt to mitigate the effects of extreme weather, take matters into our own hands and build resilience to face an uncertain future.

I was 15 when the Intergovernmental Panel on Climate Change (IPCC) was formed to collate and assess evidence on global overheating and a year later our prescient Prime Minister Margaret Thatcher – possessor of a chemistry degree – warned in a speech to the UN that:

"We are seeing a vast increase in the amount of carbon dioxide reaching the atmosphere... The result is that change in future is likely to be more fundamental and more widespread than anything we have known hitherto."[44]

She called for a global treaty on climate.

Our King, Charles III, said in 2019: *"I am firmly of the view that the next 18 months will decide our ability to keep climate change to survivable levels and to restore nature to the equilibrium we need for our survival."*[45] Five years after this statement, little has changed.

Despite the global treaties, and the pledges and the promises, the fundamental change Thatcher warned us of, is now here; our governments have not saved us and are not coming to save us. The UNFCCC COP system is a failure.

My King, my Parliament, my local Council have all declared it an 'Emergency'. The United Nations have declared code red for humanity. On 6 September 2023, UN Secretary-General António Guterres declared that: 'Climate breakdown has begun'.[46] I am left with the conclusion that *our political system is not equipped to rise to this challenge*.

When you hear that the United Nations have declared code red for humanity, that there is 'no credible pathway to 1.5°C' and when you witness the continued failure of our governments at COP, it's easy to be stunned into inaction. So, what do we *do*? How can we act, in concert, with agency to mitigate and adapt to our reality? How do we transform our future?

It's down to us. To you and I. You cannot have a million without one; what *you* do – what *we* do – right now matters. This is about ourselves, in our communities, our workplaces, our places of worship, our schools and Anchor Institutions; working together to transform our future and adapt to the change that is here.

Because the 'solutions' are available, we just need to take *collective action* and implement them in our communities.[47]

Food and water shortages are here; this is most obvious in the Majority World, where millions are starving to death or suffering systemic malnutrition, where Day Zero has already hit some cities. In the developed world, we're lucky that we've been insulated from the effects somewhat – though London has been as close as a few weeks to running out of water and we've seen rationing and lack of food (one lettuce each, empty shelves, no tomatoes). These are symptoms not only of just-in-time supply chains, but of the impact of climate decline in regions that grow our food.

Where does your food come from? Where does your water come from? How do you power your home and your transport? These are questions you, your family and your community should be thinking about right *now*. Before you are thirsty or hungry.

You'll likely be familiar with 'warm spaces', where people that are cold and cannot pay to heat their homes gather. What about 'cool spaces', where you can shelter from the heat (as recently suggested by County Councils in East Anglia and in place in cities like Paris[48] already)?

Air conditioning is a luxury we will not be able to afford, as it only adds to the climate predicament.

Carl Sagan said, *"Anything else you're interested in isn't going to happen, if you can't breathe the air and drink the water. Don't sit this one out. Do something. You are, by accident of fate, alive at an absolutely critical moment in the history of our planet."*[49]

I always thought somebody should do something about that, then I realised I *am* somebody... and if the government isn't coming to save us, then we must take care of people and the planet into our own hands to adapt to our new reality. To save everything we can that we love and cherish.

So, what's *your* plan? Find the others. Gather, vision, plan and enact. Together, be the change you want to see in the world.

Community Climate Action: How Communities Can Change the World

It is our civic duty to act. To grow community resilience in the face of crisis, by way of mutual aid. Insulate Britain were right about what needs to be done, whether or not their tactics were right; so, let's get on with insulating our homes and insulate Britain. But not just insulation and the retrofitting of our houses – *Community Climate Action*[50] can deliver renewable community energy, increase biodiversity, resilient food systems, clean water and the institutions, expertise, funding and finance to deliver, in your community. By the people, for the people, via community owned and democratically controlled institutions. This is systemic change, based on Community Benefit Societies.

Acting together, we can accrue the financial benefits of power generation and energy supply for the benefit of our communities, tackling the systemic inequality of power companies making record profits. Generating profit with a purpose – to create the environments we wish to live in and healing our planet.

Our farmers are part of our community and, with the aid of a plan, are potentially ready and able to transition to agroecology and organic farming; ensuring nutrient-dense food is available. But we must nurture these relationships with our food and our farmers, or we have no chance of bringing them onside; bear in mind that they are as much a victim of the system as others – squeezed between well-financed big agriculture and supermarkets. Even some of the best examples of 'nature friendly farming', which hugely helps biodiversity to increase, are still growing monocrops for Nestlé and others that become processed, sugar-rich products in our supermarkets. By connecting directly with our food and growers, as part of our communities, we can effect change; be it urban, peri-urban or rural. The UN has said only small farmers and agroecology can feed the world.[51] Institutions like the Suffolk Farming and Wildlife Group[52] are listening and launching initiatives to transform our food systems and launching Project Drawdown[53] in Suffolk (how we sequester carbon and return our home to a place of safety). Learn where your food comes from and how to cultivate food, how to become part of Community Supported Agriculture.

We must also work to educate our communities about climate damage and the importance of taking action. Hosting workshops and events, providing resources and information, and working with schools and other community organisations. By raising awareness about the issue, our communities can inspire others to take action and create a ripple effect of positive change. The powerful thing about raising awareness by way of setting out the need for (transformative forms of) adaptation and of resilience-building, preparedness, is that doing so wakes people up more effectively than anything else.

Our communities can advocate for policies and regulations that support climate action of all effective kinds. This can involve lobbying local, county, and national governments to adopt policies that reduce greenhouse gas emissions and support renewable energy development. By working together, communities can exert pressure on decision-makers and help drive change at a larger scale.

But why wait for a government that is dragging its heels? Let's *lead* on the change that is here and answer the question our children and future generations will ask "what did you do?". Let's create the future we wish to see, starting now.

Community Climate Action is critical in the struggle against dangerous human made climate change. By coming together and taking action, communities can reduce their carbon footprint, adapt to the impacts of climate decline, and inspire others to follow their lead.

There are many great communities and groups already taking action. Global regenerative communities, community energy groups and installations, Community Benefit Societies set up to retrofit homes. Urban and Rural movements for organic food production, NGOs acting to protect nature and a move towards participatory politics, with a Peoples Plan for Nature. We must amplify and connect this activity, coordinate action and not act in silos. A mass mobilisation of civil society is required, if we are to adapt in a way that could enable us to flourish.

What's your plan? What does your future look like?

Join or set up a Community Climate Action group in your community; Transform and Adapt:

- Map your stakeholders.
- Agree your shared values (there is more that connects than divides; seek to bring in members of your community who do not seem like your natural bedfellows).
- Think about and document what health and wellbeing feels like for you and your community.
- Write your plan.
- Form groups to tackle and deliver on the changes we need to see; transport, housing, energy, food (and water), biodiversity.
- Make it *your mission* to deliver your plan.

Again written specifically for this book, Morgan draws on his background in education to examine how TrAd intersects with the growing movement and practice of Transformative Learning.

Chapter 9

I Don't Want You to be 'More Like Me', I Want You to be 'More Like You'

Morgan Phillips

An accusation often thrown at educators who want to educate for something (like 'the environment', or 'sustainable development', or *TrAd*), is that we're arrogantly trying to turn those we engage with into ourselves. We stand accused of telling those learners we find in front of us: 'I want *you* to be more like *me.*'

This accusation is hurled hardest when – as educators – we are explicit about the values we hope to nurture through our teaching. I believe in being explicit about this, I believe in transparency, and I am extremely wary of anyone who claims to educate in a values-neutral way – it is nearly impossible to do so. Whatever you are teaching, whatever style or approach you adopt, you will inevitably be activating and reinforcing some values more than others. You may not be fully aware of this, but it will be happening.

For me, the values I aim to nurture are the compassionate and creative ones. They can be found in the Stimulation, Self-direction, Universalism, and Benevolence segments of this map, overleaf.

As well as consciously and explicitly trying to activate and reinforce these values, I also take care to not activate and reinforce the values that oppose them (the ones – roughly speaking – in the bottom half of the map, most especially those at the *very bottom* of the bottom half).

I do this because it is known, from the research, that when people hold compassionate values – those in the top right of the map – as important to them, they are more likely to want to act for the environment, and for social justice. It is also known that when they hold opposing values, those associated with 'achievement' and 'power' they are less likely to commit themselves to causes that are outside of their immediate, narrow, self-interest.

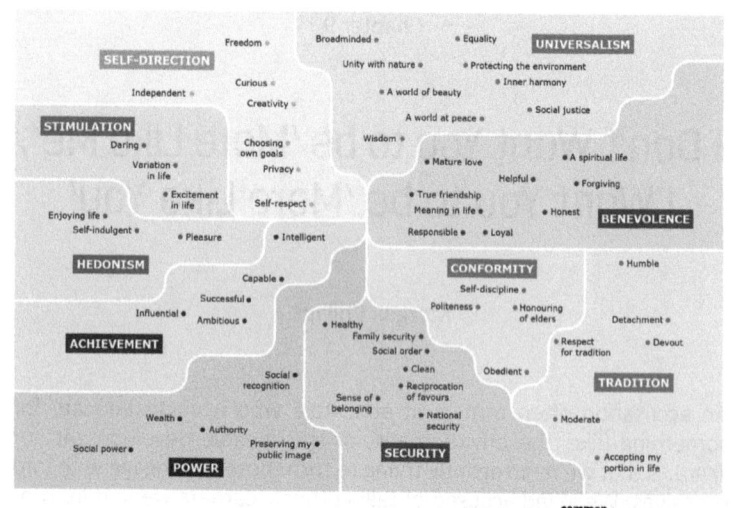

Redrawn with permission for Common Cause from Schwartz, S.H. (2006).

A similar thing is true of the north-west to south-east axis on the map. Compared to those who are obedient, detached, moderate, and tied to securing and shoring up the status quo, those who are daring, creative, and autonomous, are far more likely and *more equipped* to do the doing of something like TrAd.

The bad news is that, according to the Common Cause Foundation,[54] 77% of UK adults assume their fellow citizens to be predominantly self-interested. The 77% might be right.

It appears that, for TrAd to succeed, we need billions of selfish people to (quickly) transform into compassionate, caring, collaborative, and creative TrAdders. In short, using education and any other tool we can get hold of, we need to change billions of people's values. This is easy to say, but incredibly hard to do.

Values are very deeply held, far more deeply held than opinions, perceptions, attitudes, and even beliefs. It is not easy to change a person's values. And anyway, very few people want their values, or their children's values, to be changed. If someone stood in front of me and told me they were trying to change mine, it would feel very invasive, I would be very resistant. Ethically, it is a highly dubious mission.

So, if you – regardless of the values you hold – agree with 77% who believe that most people hold *bottom half of the map* values, and you hear me saying that we need the opposite to be true, i.e. for most people to hold *top half of the map* values, you are going to assume that I am trying to do the impossible: change the values of billions of people. This isn't, however, what I, and many other educators in this space are trying to do.

That the bleak assumption that most people are self-interested, self-absorbed even, is so widely held is unsurprising. Here in the UK, we are surrounded by a media and social media that (for profit) repeatedly highlights our selfish sides. This feeds and reinforces the idea that being selfish is a natural phenomenon borne out of a 'survival of the fittest' mentality.

If you go along with this assumption, it is perfectly natural to feel pessimistic about our fellow humans. It is hard to imagine them as anything other than fame hungry, self-serving consumers. This can get you down, it is easy to feel lonely, disempowered, despondent, and fatalistic. These are not emotions that are going to motivate anyone to build a TrAd future.

But what if this gloomy assumption about human nature is widely held, but wrong? The second half of the same Common Cause Foundation study found that it — most probably — is. Further studies are confirming this, books by Rutger Bregman,[55] Jon Alexander,[56] Ece Temelkuran[57] and others, and programmes like the Dirt Is Good Project[58] are helping to breakdown this debilitating misperception. This is incredibly good news for TrAd, TrAdders, and the future.

I used to be amongst the 77% who carried around a negative belief about my fellow humans. I'm not anymore, my belief system has been transformed by what I learned from Common Cause, what I have read, and what I have observed as part of the team who designed and delivered the Dirt Is Good Project. This has had a profound impact on how I think about my role as an educator and activist.

I value those values in the top half of the values map; I care more about equality and autonomy than I do about preserving my public image and respecting tradition. And it is true that I want you to feel roughly the same. The thing is that, whereas I used to assume that you are selfish, I now assume the opposite.

When I go into a classroom, I go in there knowing, statistically, that three-quarters of the learners will, deep down, value compassion, caring, cooperation and creativity as much, probably more so, than me — even if they rarely get to show it. Again, this is what the research shows — the same Common Cause Foundation study showed that 75% of adults hold compassionate values strongly, i.e., most people. 'Most people, deep down, are decent', as Rutger Bregman puts it.

My task then, as an educator, is to enable and encourage and support people to show their compassionate sides. *I don't want you to be more like me, I want you to be more like you.*

When learners are encouraged and enabled to show their compassionate, collaborative, creative sides, they reveal a truer version of themselves. And, as those around them reciprocate (which they often do), they do something very powerful. Together, they begin to deconstruct the myth that most humans are selfish and start to reconstruct something most of us know intuitively as young children, i.e. that most people are kind and generous and fair. The true story of humankind is survival of the *friendliest*, not fittest.

Turns out, most of us are caring, cooperative, creative, and kind; and the more we see this side of ourselves and each other, the more we begin to let go of that widely held, bleak (and false) assumption that most humans are selfish. That process of letting go is transformative, it doesn't so much change our values, what it changes is our perception of other people's values. We come to see that we have more allies than we thought, and crucially more allies than enemies.

Education that activates and reinforces compassionate values and shows them to be the norm, not the niche, is unifying, it is empowering, it fuels hope and motivates action. To make TrAd possible, educators need to change our perceptions of each other, and of ourselves. It is one of the doors we need to unlock to make Transformative Adaptation happen. This calls for sustainability education that is a profound departure from the dominant 'Direct Instruction Knowledge Rich' approaches that tragically mimics the disastrous schooling that prevails in so many hyper-individualist societies.

Transformative Learning for Transformative Adaptation

Sadly, across too many schools, charities, and even universities, sustainability education is still either absent, or basic and superficial. Other than tiny pockets of good practice, for most people sustainability education is stuck in first gear. In schools, in England especially, young people are taught a scattering of facts about nature and climate change, and then left to figure out for themselves what they can do about it. It is a 'transmissive' approach, when what's needed is a 'transformative' approach.

Transformative learning has its roots in the 1970's work of Jack Mezirow who defined it as *"an orientation which holds that the way learners interpret and reinterpret their sense experience is central to making meaning and hence learning"*. Put more simply, it is the idea that learners don't just acquire knowledge, memorise it, and then repeat it when asked to (for example in a test). Learners instead, think critically about what they're learning as they learn it. They are using their new knowledge to test previously held beliefs, perceptions, and worldviews, which often leads to a change, or a transformation in those beliefs, perceptions, and worldviews.

For me, the revelation that most people, deep down, are decent, was transformative. It changed the way I look at other people, it changed the way I approach my work as an environmentalist, it changed who I am as a person. It has helped me to believe that TrAd is possible. It has also helped me to acknowledge other ways in which I have been transformed. The list of beliefs, perceptions, and worldviews I've held and then let go of is longer than I care to admit. I feel differently now about race, gender, sexuality, meritocracy, inequality, colonialism, politics, monarchy... I owe a huge debt of gratitude to the many educators I've encountered throughout my life. And I am still being transformed.

TrAd needs TrAdders, a critical mass of us who are active, resilient, and courageous. There is a huge number of potential TrAdders out there in the world, but they are not TrAdders yet. To be a TrAdder in modern Western 'civilisation' is to be an anomaly. To transform into one requires one to let go of beliefs, worldviews, and perceptions, and to form new ones. Depending on your starting point, this can involve very many mini-transformations, and/or one major transformation. Transformative learning interventions, carefully designed and skilfully delivered, can facilitate this process. Without them, we may never see the emergence of Transformative Adaptation.

In this chapter, bestselling novelist, Manda Scott, describes how we can break through the current destructive capitalist system and design a flourishing future for generations to come. Her 'Thrutopian' vision is a transformative adaptation at the vital level of scenario-planning and creative, reality-based thinking and storytelling.

Manda argues that at present all of us, to a greater or lesser extent, are locked in the heart of a death cult that insists there is no alternative, while alternatives quite evidently abound. What we lack is a shared vision of a flourishing future and to find one, we need stories of how our lives would look and feel if we let go of our encultured drive to engage in a market of goods and services, and instead became just as curious about – and invested in – caring for the entire web of life, of which humanity is so clearly a part. As Chris Smaje succinctly framed it, "We are already a keystone species. We need to find ways to become a good one."[59]

This chapter draws together material from a series of articles by Manda in **Permaculture** *magazine. We hope and think you will find it a great way of exemplifying what TrAd is or ought to be all about.*

Chapter 10

Thrutopia

Creating a New Story for a World Undergoing Transformation

Manda Scott

We live in capitalism, its power seems inescapable – but then, so did the divine right of kings. Any human power can be resisted and changed by human beings. Resistance and change often begin in art. Very often in our art, the art of words.

Ursula Le Guin,
speech at the National Book Awards,
20th November 2014[60]

What do you think the world will be like ten years from now? Or twenty, fifty, a hundred, a thousand? If you close your eyes and really imagine yourself waking up one sun-drenched morning in 2054, how do you feel?

If the answer is relaxed, optimistic and joyful, then you are one of the few who has divined a way forward to a flourishing future we'd all be proud to leave to the generations that come after us.

I suspect many readers of this book do fit into this category: loving the land and our communities is how we live, and by now most of us know that the continuation of neoliberal/free-market/extractionist/predatory/whatever-we-want-to-call-it capitalism is not compatible with a flourishing web of life – or even its continued survival.

We are in the minority, though. Most people currently struggling through the days in our crazy world still find it easier to imagine the total extinction of life on earth than to imagine – and so make credible plans to achieve – an end to predatory capitalism and the entire political, economic, business, social and cultural death cult that is built in, on and through it.

None of which is surprising when the vast majority of adverts, newspaper headlines, popular track lyrics, TikTok videos, TED Talks, vlogs, political policies, TV binge sets, novels, biographies, memoirs and documentaries in print and on screen, continue to push a world view in which business is usual, economic growth is a worthwhile end that justifies any means, and success is defined by amassing money and the power that it confers. We live in a world that tells us scarcity, separation and powerlessness are the way it has to be,[61] and we believe it.

What would it take to turn the super tanker of our cultural narrative away from the cliff's edge? It's not, after all, as if we're lacking coherent ideas of where the current system is taking us: every second Booker prize nominee marries *A Handmaid's Tale* to *The Road* and gives birth to the living hell where those with the biggest baseball bats take total control of the starving survivors.

This is not a useful narrative, not because it's untrue (it isn't), but because it's not effective. Scientists have been sounding increasingly strident alarms about the implications of rising CO_2 levels for decades and yet we're burning more fossil fuels than ever before. Petroleum geologist and energy analyst Arthur Berman says that over half of the oil ever burned has been consumed since 1995[62] and we can't pretend we didn't know it was a bad idea.

Terrifying people into changing behaviour has never yet worked: our brains are just not wired to respond to hypothetical scare stories that don't have immediately obvious links to the present. 'Things will be bad in a decade if we let the oceans die,' has no traction in a reality where people are swimming in blue seas that still give up fish, while shifting baselines mean we don't recognise that the current fish populations are a fraction of what they were and raw sewage has not always been a standard surfing hazard.

Doing the same thing time after time and expecting a different result is the very definition of madness. Or at least, of ineffective action. But to find something effective, we need to explore why we're not changing. We are, after all, an astonishingly, spectacularly, brilliantly creative species. When we're given a problem, we can produce the most remarkable ways to circumvent it. I could write you pages of proof, but actually, you just need to flick through the rest of this book, or any issue of *Permaculture* magazine, or to access one of the host of really creative podcasts that explore those at the leading edge of change.

The Dis-imagination Machine

So why are we hovering on the edge of a collective cliff? Core to this has to be the concept that predatory capitalism is a dis-imagination machine,[63] designed to narrow the range of our creativity. When our autonomic nervous systems are in sympathetic overload, we focus on survival, not on the beauty of the sunset, or the myriad ways we could solve the meta-crisis. And, by accident or design, we're surrounded by highly visible signifiers of 'failure' to thrive within the system, the most horribly obvious of which are the now-ubiquitous homeless men and women, huddled and freezing on the pavements of even our smallest towns.

I grew up near Glasgow and homelessness was not a thing in my childhood. Shifting baselines turned this, too, into a new normal — until the global pandemic forced the government to act and we discovered that providing everyone with a bed for the night was not hard and cost a tiny fraction of the contracts doled out to ministers' friends to provide ineffective PPE and double count the gloves.

With the daily threat-reminder removed, we began to realise that we weren't, in fact, born to pay bills and die. The Great Resignation saw a great many of us throw in our bullshit jobs[64] and find better ways to live... until the wheels turned, fossil fuel prices soared (bear in mind prices were rising before a large petro-state invaded its neighbour)[65] and suddenly the core needs of food, warmth and shelter became scarce again.

So, what can we do? Basic neuroscience says that any organism responds better to positive reinforcement than to punishment, but also that there's innate reinforcement in staying still: absent emotional, physical or spiritual violence, the known generally feels safer than the unknown.

We need powerful incentives to move beyond our comfort zones, and throughout human history, the thing that would reliably lure us out into the unknown was a good story. Dick Whittington didn't pull on his Puss-Boots and stride forth for London because life was bad at home, even if it was: he went because he let himself believe the streets were paved with gold.

Closer to our time and place, we all know that nobody puts their family in a tiny boat adrift on turbulent seas in the depths of winter just because their current existence is intolerable; they do it because they believe life will be at least survivable when they land. Every single active choice we make in life is born of a dream telling us how life will be better when we've done the thing, whatever it is. Quite often the new home, job, partner, car, child, puppy, does not, in fact, make us happier, but we exist in a perpetual cycle of hope over experience that drives us on to the next gold-plated promise.

An Ecosystem of Stories

So, if our current culture is crumbling in real time, we need not just one story of how the world could be better, but an entire ecosystem of stories showing how a flourishing system would look and feel from the inside. Crucially, we need also to see how to get there in ways that not only feel doable, but desirable. When we've got at least the core concepts of this sorted, we need to get anyone and everyone talking about it in every medium – from supermarket conversations to tabloid op-eds to TikTok videos – until we reach a tipping point where a media system that is dedicated to maintaining the status quo has to listen.

This may feel like solo-ing El Cap[66] but there are at least a few handholds and bolts on the route. First thing to take on board is that tipping points in the narrative do happen. There was a time when everyone on television smoked. These days, characters only smoke to provide highly specific (cliched and narrow) cultural signifiers. Change happened swiftly and effectively when the cultural narrative surrounding smoking changed.

The corollary to this is that weaning some people off a nicotine addiction is a walk in the park compared to weaning an entire culture off its addiction to fossil fuels. And yes, we have achieved other cultural tipping points. But ending slavery (insofar as we have done) or sexism (insofar as we have done) or racism, or homophobia or [fill in the blank of your favourite cultural shift] all took place within the envelope of the extractivist system. They changed the franchise. They did not seek to overturn the absolute baselines of the system itself, which is what we need now. So we can take comfort from the fact that tipping points happen and acknowledge that we have no idea what it will take to create one of the magnitude that we need, only that it'll need all of us and time is short.

A New Thrutopian Genre

All of this is why some of us are endeavouring to create a new Thrutopian genre. Named with the permission of Emeritus Professor Rupert Read after his 2017 paper,[67] 'Thrutopian Narratives', by definition, map out detailed, constructive pathways that walk us forward, step by plausible, desirable, obvious step from exactly where we are now to a flourishing future we'd be proud to leave to the generations that come after us.

I can't say often enough that this is not just a job for writers. We are all going to live in the world we build, so it follows that as many of us as humanly possible need to be a part of imagining how it could look if we get things right. We need engineers and educators, chemists and crafters, growers, makers and builders, thinkers and planners and people who can work out how to craft a functioning democracy that brings out the best in everyone. Above all, we need to abandon right vs left, tabloid vs broadsheet, Twitter vs Mastodon, and find the deep wellspring of common values that join us as humanity working in concert with the web of life.

Inevitably, some of us are generalists and some specialists. The latter can delve right into the heart of their chosen subject: food systems, economic and monetary systems, power generation and how we reduce our consumption, reshaping democracy, reshaping the commons, particularly the attention-commons that is powering through the thermocline of trust[68] so fast that it might take our capacity for cohesion with it.

With all this as a foundation, the generalists can gather up the specifics and weave of them a net that builds towards a new system. Then, when we've begun to build a scaffolding of what the next steps look like, we absolutely will need writers and creatives to stop writing in the old paradigm and start writing in the new.

This, too, can't be said too often. We need to let go of the old. If we're writing something — anything — from a recipe blog to a blockbuster Netflix super-binge, we need to be writing it in the frames of the world that's coming, not the one we're leaving behind. When publishers and film-makers suddenly discover they can't get any of the old costume dramas, or super-hero fantasies based in the old power structures, but there are, equally suddenly, a whole plethora of interesting, engaging, dramatically brilliant scripts and stories built on a new set of values and assumptions — then the avalanche will start.

So this is our proposal: for each of us to bring ourselves into a generative space from which to live, dream and communicate and start mapping out routes to a future where our grandchildren's grandchildren's grandchildren look back at our efforts and congratulate us.

Each of us has an integral part to play in this weaving of possibility so that the futures we build will be generative for all life. The broad concepts are relatively easy, but having recently abandoned the writing of historical fiction in favour of writing a wholeheartedly Thrutopian novel, I can testify that crafting the granular detail of exactly how we could get from where we are to where we want to go in ways that feel plausible, desirable et cetera is a lot harder than it might seem.

Apart from anything else, this is an emergent process heading for total systemic change and, by definition, we can't predict how the new system will look any more than the first caterpillar could predict the first butterfly.

But just because we don't know the answers doesn't mean we shouldn't be asking the questions, and we already have a pretty good idea of the conditions that will take us as close to the leading edge of our current system as we can get.

The Story So Far – and Co-Imagining the Future

Can we redesign our social and political systems so they genuinely represent people and Nature, so that the future will be generative for all life? To begin to approach this question, consider the following quote:

> "When future generations look back upon the Great Derangement, they will certainly blame the leaders and politicians of this time for their failure to address the climate crisis. But they may well hold artists and writers to be equally culpable – for the imagining of possibilities is not, after all, the job of politicians and bureaucrats."
>
> **Amitav Ghosh,**
> *The Great Derangement*

So far in this chapter we have outlined the need for a new kind of narrative for our times, one that would map in granular detail the route from exactly where we are to a flourishing, generative future we'd be proud to leave to the generations that come after us. We've found that the old ways of frightening people into change haven't worked, indeed, there was never any chance that they would. But we've also established that we are a storied species, and while we're alive and an integral part of the web of life, then there's a chance that we can move away from the death cult of predatory capitalism and orient instead towards wholeness. In order to power this shift, we'll need an unstoppable wave of stories to fire the collective imagination until a different future seems not only possible, but inevitable, and everyone wants a piece of it. Stories of the kind that Amitav Ghosh called for. I suggested earlier that we need to create an entirely new narrative genre in which to wrap these stories and called it Thrutopian, after a paper by Rupert Read, one of the co-founders of the Climate Majority Project, as explained in earlier chapters in this book.

And, while Amitav Ghosh is absolutely right that our current crop of politicians and bureaucrats is wholly unsuited to the task of imagining generative futures, we need to be clear that it's not only those who identify as artists and writers who are going to be part of this evolutionary Thrutopian surge: everyone who can imagine a future built on cooperation and integration with the web of life will be an integral part of this new weaving. All change has to be consensual and peacefully enacted so we must all have a hand in building it.

The task now therefore, is to map out the key components of this new, emergent system. What follows is my best guess at some of the essential moving parts.

Community

The first, biggest and most frequently stated requirement as we build towards a flourishing future is that we engage wholeheartedly in communities of place, purpose and passion.

We denizens of the twenty-first century are graced with Palaeolithic emotions, mediaeval institutions and the technology of gods[69] and this, it turns out, is a fairly volatile combination. Nonetheless, it's who we are. For several hundred thousand years, we lived our entire lives surrounded by people who cared about us and about whom we cared. We garnered our sense of being and belonging in a complex, dynamic web of relationships. We found pride in bringing the best of ourselves to the collective table, gaining and giving respect, care and wisdom.[70]

Predatory capitalism has done its best to sever our connections to each other and the web of life while offering escalating addictions to patch over the wounds of our loss. The damage has been generations in the making, though the extraordinary expression of isolationism and individuation has seen its apogee in our time.

Nobody is suggesting that our Thrutopias take us back to our forager-hunter tribal inheritance, but we can undoubtedly move forward to a world where the transient dopamine hits of our mis-named social media platforms are replaced by the resilient, stronger, enduring serotonin meshes of community.

We exist at a point of hyper-complex communication systems and so our communities need to reflect all the scales on which we live.

Bringing together values of connection, compassion and cohesion, we can, and must, cultivate the capacity both to see ourselves as citizens of the whole earth and at the same time, to engage at the smallest viable scale of kith, kin and village-scale community. Ideas of how these can work abound in the Transformative Adaptation field, in permaculture and on the web at large. Engaging with them is our first, deepest and most profound priority.

Democratic Renewal

So far, so obvious. But creating local food webs, or inter-national communities of service hasn't yet changed the trajectory of our culture, or significantly dented the meta-narrative. For us to create deep and lasting change – for people to see that the old ways are dying and that genuine change is possible – we need our second key component of our newly refurbished system: fundamental DEMOCRATIC RENEWAL.

The purpose of any healthy democratic system is to create a fair, transparent means of enabling the brightest and best hearts and minds of any generation to engage most effectively with the problems of the time. The current system fails on all counts and is no longer fit for purpose, if it ever was. Party politics is corrupt, wholly owned by the corporate lobby and offers no alternatives to predatory capitalism.

Because our evolution must be peaceful, we'll need at least two steps to transition away from the existing model. The first of these is to 'fork the government' as happened in Taiwan in March 2014.[71] We can't replicate the detail exactly, but at local, national and international levels, we can elect parallel governments capable of dealing with the same issues as the ones currently holding power, but which demonstrate how a flexible group of individuals unhampered by ideology could work.

These parallel elections have a number of core components, first of which is the removal of the Party system. Coalitions of individual candidates could consolidate around shared values and agreements of how to solve problems per the Flat Pack Democracy model instigated in Frome,[72] but the vote-buying lists of the manifestos have to go. This is not about a group of people telling us what to do. It's about finding the people who can listen and engage with each other to problem-solve. The problems themselves are sourced in Citizens' Assemblies convened with rolling membership at each level of governance.

The second essential component to our new democracy is to shift the voting system away from First Past the Post towards weighted, time-dependent proportional representation in a form that allows both for nuanced expressions of opinion and absolute transparency. This area is evolving fast and I suspect my ideas of what can work may have changed between my writing and your reading of this, but just now, I'd go for a combination of quadratic[73] and conviction[74] voting on the blockchain.[75]

Voting would be compulsory, as it is in Australia, but instead of allowing voters to spoil the ballot, there would be a 'NOTA' (none of the above) option on the paper, with the proviso that if NOTA wins, then the election in that constituency is held again and not one of the candidates on the failed ballot is allowed to stand.

With our candidates chosen and our voting in the bag, we'll need a transparent, reliable means of registering the votes cast. Paper ballots are fine as long as they are collected, collated and counted properly, but there are electronic means of achieving total transparency and freedom from interference, again linked to the blockchain.[76]

It won't take long for our parallel governments to demonstrate that they function far better than the current mess. It will take longer to persuade the media ecosystem that thrives in and through them, so we'll need, along the way, to generate our own means of connecting and exchanging ideas in ways that promote cohesion and not division: again, Taiwan is leading the way.[77] Abandoning newspapers, legacy media channels and toxic social media outlets is a key part of the new paradigm and something we can all do today. (Instead, listen to podcasts. Your worlds will open up.)

We still (just) have government by consent. When a critical mass of the population understands that the new system is better than the old, then the new one becomes the chosen form of governance.

With this in place, our second step is to re-localise decision making, building participatory governance functioning from the ground up, with power left in the hands of the people nearest to which it will be used.

Thus, our communities of place, passion and purpose come together and work out how to identify the smallest sane subunits within which decisions can safely be made without coercion or corruption. Numbers around a dozen allow for all voices to be heard in each electoral subunit, while the collation of these into a functional governance area needs around the Dunbar number of approximately 150.

Thus, a village/parish/street/zone of around 150 people congregates in person or online and is split into subunits of around a dozen in ways that guarantee households are not kept together (drawing cards from a well shuffled pack works, or random number generators online).

Each group of twelve elects one or two speakers from amongst their number to go forward to the governance unit. If necessary, facilitators from outside the zone come in to ensure that the loudest/most charismatic/most frightening voices don't dominate. We need the brightest and the best, not the kleptomaniac sociopaths the current system elevates.

Those elected form the lowest rung of the governance structure and are delegated the tasks of the local area, but note again, the nature of the tasks are defined by a Citizens' Assembly drawn at random from local participants. The delegates are there to get things done, not to define which things need to be done. If matters arise that they believe need to be dealt with, they send them back to the Citizens' Assemblies.

Additionally one or more of this core governance group are chosen (either at random or by vote: decision to remain with the divisional local entity) to go to the next highest governance zone, from parish council to town council, say, with the proviso that at every level the discussions are transparent and if the electors feel they are not being well / accurately represented, they have the right of recall of their delegate.

This iterates on up all the way to national government, but even at this level, the individuals thus elected are not the executive arm of government. They are, instead, a group who bring the best of their administrative abilities to bear on the implementation of tasks that are defined by Citizens' Assemblies with rolling six or twelve month memberships. (That's to say, a handful turn over every few months so that fresh viewpoints are always being brought to bear.) The Citizens' Assemblies are facilitated in ways that ensure all voices are heard and that promote cohesion, not division.

All of this combined brings us to a means of governance where human creativity has an opportunity to emerge at a local, national and international level. To escape entirely from the current model, we'll need a new means of sharing, exchanging, storing and accounting for value. We currently call this the economy. The death cult economy is designed to grow, regardless of the harm to people and planet. What we need now is a system designed to meet the needs of all people (all life) within the means of the living planet.[78]

Permacultures of Value, Needs and Wants

So we come to the central question of how a regenerative economy might look. We'll see how that leads us to a place where we can and must examine in detail what we need in our new communities and how we can employ the logic of degrowth to reduce our power consumption to something approaching a regenerative scale. In short, our topic now is economic transformation – and therefore why we need to create positive narratives to drastically shift us towards collaboration, agency and sufficiency. As in the following powerful remarks, all the more emotionally resonant for coming from youth voices:

> *"Imagination is the soil that brings dreams to life.*
>
> *It enables us to bring clarity to what we want to change and what is truly possible. Unfortunately, research shows that as we go through school, our ability to imagine decreases steadily.*
>
> *We need to recover the ability to imagine if we are to create the world of our longing. [We can... prefigure] the kind of world we wish to create through the way we show up with one another and the spaces we create."*
>
> **YouthxYouth,**
> Youth education activists[79]

In this chapter I've acknowledged that working out the exact steps forward and weaving them into a contiguous whole is hard, and that we'd need to establish some of the component parts. I therefore suggested changes to the cultural and democratic governance systems that would allow the best and brightest to move into agency at local and national levels and suggested that this laid the groundwork for shifts in the economic model that would allow us to address the core of our current malaise: our addiction to fossil fuels and the destruction of the web of life that arises therefrom.

Economic Transformation

So, let's dive straight into this need: for a total re-shaping of the aims and functions of our global economy. This targets the heart of the giant vampire squid wrapped around the face of humanity with ways to peel its tentacles off ourselves one by one, allowing us to find a new way to connect our needs, wants and creations. The whole of the 'Masters in Regenerative Economics' at Schumacher College is devoted to this (and, full disclosure, I'm an alumnus of that course) so this will be a massive abbreviation of all that's required. The point is to offer a scaffold on which you can build change into your visions of the future: it's never the finished product.

In creating new ways of using money, we need to understand how it functions in the current model. In particular, we need to understand that the hackneyed fiction in which we all pay our tax dollars/pounds/euros to an impoverished government whose capacity to pay is constrained by the tax income, is at best a misunderstanding and at worst a deliberate lie. I'm inclined to believe the latter but Positive Money established, in 2014, that 85% of MPs in the UK's parliament didn't actually understand where money came from and it doesn't seem to me that the average intellectual capacity has increased, so perhaps it's just idiocy.

To be clear, money in our current society is made up out of nothing and then sold to us at a profit by the people who are given the power to do this. Those people become very, very, very rich. We call them bankers, but really, they're just makers-up-of-money who then gamble with it in an astonishingly complex pyramid scheme where they get to tell the world that their fictional money is worth more than it used to be.

As far as the money we actually see goes, there are two basic sources: banks lend money into existence and governments spend money into existence. Taxation traditionally has two functions: to even out the playing field (we can assume that function has lately been abandoned) and to prevent the economy from overheating (though if inflation happens as a result of material price rises without the control of the government, then raising interest rates and/or increasing taxes is not going to bring it back down. Someone might like to point this out to the Bank of England).

Note, though, that taxation is not the sole means of financing social and community goods. In fact, when the government takes in tax (or repays the national debt) it is removing this money from circulation. Forever.

The choice to link taxes-in to spending-out is political and has been since at least 1694 when the Bank of England was set up, though I would argue that Nero's devaluing of the Denarius means even the Romans were cooking the treasury books and I doubt if he was the first to think up ways to fudge the numbers. Power corrupts, and all that...

None of this is surprising. At its base, money is an agreement we all make about how to exchange, share and account for value, but it is also intrinsically linked to hierarchies of power. Fiat currencies – that is, the ones we all agree have value (as opposed to Monopoly money, or currencies issued by countries we don't like) – are always backed by violence. Just as a nation state is the geopolitical entity which grants itself the monopoly of legal violence within set boundaries, so too is a fiat currency that article of accounting issued by a nexus of power which has sufficient violence to impose its use.

Origins of Coinage

If we take it as read that Graeber[80] is right and money did not arise out of barter, but instead out of a hugely complex web of social connections, then coinage arose out of the fact that one unit of the hierarchy (a person or group) had the power to impose its value on the local area.

'This bit of silver with Our Bloke's face on it is worth a hundred times more than the bit of silver you have in the box in the corner, and no, you can't put your own face on that bit, or make one that looks just like Our Bloke because we'll kill you and enslave your children if you do,' is a conversation colonisers have imposed on the colonised for millennia. Addiction to power, not money, is the root of all evil, but money both grows out of, and cements, that power. It's telling that when the self-proclaimed Masters of the Universe summoned Douglas Rushkoff to a luxury hotel in the middle of nowhere to ask him how to manage The Event, it was the breakdown in dollar value that worried them.[81] If your power is based on the number of dollars in your multiple offshore accounts and dollars stop being worth anything, how do you control the small army of security guards you've hired to keep you safe?

A World Without Fiat Currencies

These are not problems that bother most of us, but nonetheless, it's time we started to explore what a world without fiat currencies might look like. Or one in which fiat currencies gain their value without the abuse of power.

As with the shift in democracies, we have to start where we are and aim for a trajectory that doesn't leave ordinary people destitute. So the questions become: What can we safely use to exchange value (what currency(s) can we all agree on)? How will we distribute this value-unit? What will we be paying for? And to whom?

For the first of these, physical currencies have the advantage that nobody is checking what we spend and where, but digital currencies can be created that are less easy to forge and that get round their own intrinsic 'double spend' weaknesses. (If I give you 5 Solar Coins for some power that came off your roof to my washing machine, what's to stop me giving the same SC5 to my neighbour on the other side in exchange for some baby-sitting?).

Bitcoin is a disaster in too many ways to list here (though the use of power is a huge one), but the concept of using the blockchain cryptocurrencies as a transparent, autonomous distributed ledger, free from forgery, double-spending or retrospective change is attractive and there are many new forms of currency taking off. The problem will be stopping them becoming Ponzi schemes, which will require a rapid transition away from the current system: anything that can be counted in dollars is going to be ripped off; anything that can't be counted in dollars is going to have a very limited circulation.

To get away from the grasping squid-tentacles, we need to look at value distribution. In the short term, while we're still in the old currencies, we need to implement the proposed democratic systems to allocate level-appropriate budgeting. In the very short term, we don't even need democratic change to kick this into reality. Participatory Budgeting is already taking off in places as diverse as Scotland, Brazil, Iceland and the Philippines. When areas have democratic control of the distribution of value, they can take up the Preston model to ensure that local money has local impact. If this is done at tiered levels, then we begin to see the end of the six-continent, just-in-time supply chains that dominate our current world, and start looking at a world where local goods and services supply local people, and the things that we move freely across the planet are information, enthusiasm and ideas of best practice.

Overshoot

The next question is 'what are we paying for?'. This reaches right to the heart of the change we need to see. Our current economic models tell us we're paying for 'goods and services' without pointing out that many of the goods are really very bad, in that they inevitably entail extraction, destruction and pollution.

We're a species in overshoot, addicted to a transient supply of abundant energy that's going to kill us if we don't wean ourselves off. We all know this, but the question is by how much and how soon. This is (another) area of intense debate, but if we assume that some of the super-brains are right and our current world energy use is a rolling 17 terrawatts[82] (where one terawatt is (a) the amount of energy the entire world used in 1890 and (b) the average energy released in the burning of five billion barrels of oil) and we need to get down to 5tW to give the web of life a chance to recover, then we need to have some serious conversations about what we need as opposed to what we're used to, or find convenient, or advertising has told us we want.

At baseline, we need clean air, clean water, warmth and shelter and healthy, nutritious food grown in a way that enhances the biosphere – and community. The rest is up for grabs.

To be blunt, a lot of companies will simply cease to exist and those that are left will operate on something akin to the Future Guardian Governance model pioneered by RiverSimple[83] in Wales, which creates six Custodians all with an equal say over the company's direction: shareholders, customers, employees, the local community, the environment and the supply chain. These companies will exist because they can offer useful local goods that are actually good and services we actually need. Importantly, the bullshit jobs will be gone, partly because nobody will be making pointless widgets, but also they won't be sitting in call centres or stuffing boxes in a warehouse for home deliveries. How we spend our time will be far more aligned to what our communities need. If we're smart (and I believe we can be), we'll find ways for everyone to explore what makes their heart sing and to bring that to the table in service of repairing the harm already done and preventing more from happening.

If we'd been smart in the past, we'd have acknowledged that fossil fuels were a global commons and used our blip of ancient sunlight to build the most amazing post-fossil infrastructure. We're a century too late for that, so instead, we need urgently to work out how best to use the final tiny fraction of the energetic bonanza on building an avenue through to the future we need, while minimising the impact of a toxic global supply chain.

The Post-carbon Narrative of Cultural Abundance

To steer our super-tanker away from the cliff's edge, we'll need to create a post-carbon, post-growth, post-urban, post-work world where our lives are spent in service to healthy, integrated communities in which we feel safe, confident and have a clear part to play. We'll have worked out what we're here for, individually and collectively, knowing that this will evolve over time, but for sure we can pass on the idea that being is more important than doing and that if each of us does what only we can do, and do that thing well, then we'll be playing our part in the greater web of life. This is our greatest gift to future generations: this and an inhabitable, flourishing, life-filled planet.

So this is it. We need stories and we can build them in ways that will lead people forward. The core components I have suggested are the ones that make most sense to me. You may have others. I hope you do. Just as there are a thousand different iterations of the early foundation myths of our planet, so there need to be a million different variants on the flourishing futures we could build and how we get there. What matters is that we're all thinking about this instead of obsessing about the ugly love child of Handmaids and Roads: where we put our energy is where we get to.

What matters is that we – all of us – create as many narratives as we can in as many forms as we can to generate an inexorable shift away from the old certainties of separation, scarcity and powerlessness to explorations of collaboration, agency and sufficiency.

You don't have to write your ideas down if that's not your thing. Speech is our most powerful medium. Word of mouth still outweighs a clever meme on the screen. Every conversation matters now, whether it's with a bank teller in your local branch explaining why you're moving your account to a bank that doesn't invest in fossil fuels, or with the local supermarket manager explaining why you're moving your custom to local store that wraps fresh goods in paper; conversations with family, friends and colleagues need to revolve less around football and more around the need for urgent political and economic change. People will move together if they can see a future they'd actually want for themselves and their children. It's up to each of us to create it.

Towards Permacultures of Intelligence and Wisdom: How Change in Complex Systems Emerges

Along the way thus far in this chapter, we have glossed over a number of ideas, notably the theory of change in complex adaptive systems: there is more to explore here. Not just because some of these ideas need to set deeper roots in our imaginations, but also because the world is changing faster than we might imagine. So let's stare a little harder at the nature of systems, tipping points and the theories of change that are centred around them. Many of our core concepts were proposed by the Nobel laureate Ilya Prigogine.[84] He was a nuclear chemist, and it may be that concepts which apply at the atomic level don't translate perfectly to the level of hyper-complex human cultures in a world of global real-time communication, but they do offer a model that we can grasp and I have yet to come across anything better.

Prigogine established that: 'When a complex system is far from equilibrium it reaches a bifurcation point where further instability leads the system either to collapse into chaos and extinction or to emerge into a new system.'[85]

Emergence

This is where our theories of emergence from complex systems arise from, and these, in turn, are the foundations on which we build hope: not the pre-Tragic[86] hope of the techno-optimists and eco-modern dystopians who see technology as paving the way to a future in which the web of life is an 'aesthetic option' (which can, by definition, be rejected and then destroyed), but a post-Tragic hope that sees the reality of our Meta-crisis[87] in all its raw chaos, but holds to the possibility of emergence as something worth striving for.

If we're going to pin our hope on this, then we need to understand a little more of what emergence looks and feels like. Happily, our world offers endless examples of complex systems that emerge into new systems under specific circumstances.

Most commonly referenced is the caterpillar, which, at a given point, spins a cocoon and dissolves into the chaotic DNA soup of a pupa. Pretty soon, imaginal cells arise amidst the chaos. In the beginning, these are assaulted as being 'other' and removed by the system, but more emerge and begin to coalesce into imaginal islands which cohere to form imaginal organs which eventually become the butterfly or moth that (in a rather neat personification of our metaphor) emerges from the cocoon – as suggested by the TrAd logo.

This leads us to Prigogine's second most quoted aphorism, which states that 'When a complex system is far from equilibrium, small islands of coherence in a sea of chaos have the capacity to lead the entire system to a higher order'.[88]

This is acquiring the weight of prophecy in certain circles now and those of us who bandy it around tend fondly to believe that we're in the process of creating small islands of coherence focused on whatever we hold most dear at the time. We may even be right.

It's a comforting thought, but one we can't really test, because one of the overriding axioms of systems thinking is that no system is predictable from the standpoint of the one that preceded it.

If ten thousand ants mill around on the edge of a chasm that none can cross alone, the exact nature of the bridge they will form with their bodies is wholly unpredictable. The fact that they can, and probably will, form such a bridge, though, is as predictable as is the emergence of the red admiral from the cocoon: it's the caterpillar or the single ant that can't predict the outcome. Those of us standing outside can make an educated guess based on previous experience.

Living, Breathing Adaptive Miracles

So let's park this and consider the current moment in the history of humanity. Each of us – you, me and anyone we meet – is the endpoint of billions of years of evolution. Our distant predecessors were hydrogen molecules and, when those hydrogen molecules gave rise to more complex atoms and then compounds and finally to life forms that could reproduce; we are the descendants of those that lived long enough to do so.

The odds against us being here are staggeringly high. We are living, breathing miracles. We are also living, breathing complex adaptive systems formed of cells that are themselves complex adaptive systems, bunched together into organs that are complex adaptive systems, to make human beings who are complex adaptive systems who then join together in communities of place, purpose or passion to form hyper-complex adaptive systems that span the globe and communicate across space in real-time.

In a healthy society our networks of place, purpose and passion would be generative spaces where we were encouraged to bring the best of ourselves to a shared endeavour whose focus emerged from the totality of the system. There are places in the world where this is happening.[89] By and large, though, in our crumbling end-stage somewhat toxic mess of a culture, our communities tend instead to become echo-chambers of self-delusion.

Linearity

Inherent to these echo-chambers is a clinging to linear thought. We are the inheritors of what was laughably called The Enlightenment, which projected the emotional illiteracy of public schoolboys who grew in an atmosphere of punishment, privation and (at least financial) privilege onto an outer world that was wildly scary and existed to be tamed.

Imposing linear narratives of cause and effect gave the illusion of control which allowed this astonishingly toxic culture to spread narratives of progress over processes of colonisation, corruption, inequity and wanton destruction while ignoring the wider systemic impacts until way, way too late.

If you're reading this book, then by now you know this. You know, too, that linearity is not the way the world works: the Law of Unintended Consequences applies to pretty much everything we do (I'm using 'we' here to denote the hegemonic narratives of the death cult, not necessarily you and me) and the ripple effects are vast, unmeasured and very likely unmeasurable.

If we pull a couple of areas at random out of the hat of unknowing, then farming and health are exuberant examples of policy areas where linearity has been practised to catastrophic effect.

Mid-way through the last century, we were pretty good at measuring nitrogen, phosphorus, potassium (NPK) and acidity (pH) levels in soil, and had 'demonstrated' to ourselves in carefully controlled trials that ignored the astonishing complexity of the soil food web that these and only these were essential to the growing of healthy plants.

We turned the creation and application of these macro-nutrients into a global industry that proceeded to destroy living soils worldwide, breaking through planetary nitrogen and phosphorus boundaries with terrifying speed[90] while wreaking havoc with freshwater and marine ecosystems and, at the same time, creating food that was and remains deficient in micronutrients and phytochemicals and almost certainly a host of other things we can't measure yet. We are now at the point where Dr Robert Lustig asserts that 93% of US citizens have metabolic disease signified by dysfunctional mitochondria.[91]

We are just beginning to understand the complex links between the soil biome and the human (and animal) gut microbiomes, but already we can safely predict that processed foods do not contain the complexity we evolved to need and that the eco-modernist concept of 'Precision Fermentation' is attractive only to the small sub-group of technophiles for whom the term was focus-group tested. The rest of us want (need) real, whole food grown in living soil as close to our location as possible.[92] Yay for permaculture.

So let's throw in another aphorism: no problem is solved from the mindset that created it.

Applying this to the theories of change in complex systems tells us that our linear cause-and-effect mindset is not ever going to be able to map a way through, that we won't solve the meta-crisis with regenerative farming, or permaculture, by all turning vegan, ending the use of fossil fuels, regulating Forever Chemicals or [fill in your favourite fix]. There is no possibility that any single thing or even suite of things will fix this: the Meta-Crisis is bigger than all of us.

So What Can We Do?

Back in 1999, Donella Meadows, one of the greatest minds in the systems thinking world, created a list of 12 leverage points that could affect change in any given complex system (see below).[93]

Meadows stacked her leverage points in order of efficacy with 'Changing constants, parameters and numbers' lowest on the list: essentially tweaking regulation, laws and tax points. She proposed this in the last year of the old millennium and one of the many distressing features of late stage capitalism is that, two decades on, the system still elevates to power people who believe that tweaking constants and numbers is a useful thing to do. Granted, the system is designed to amplify the clout of people who will fight tooth and brain-dead nail to maintain the status quo, but even so, it's hellishly depressing.

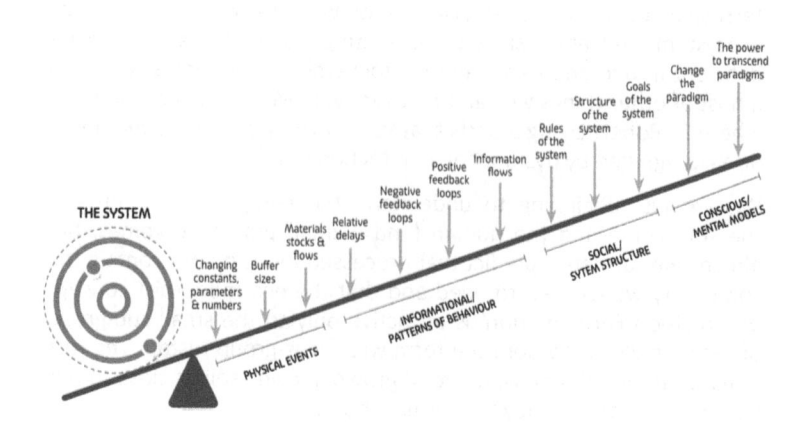

Letting Go

So let's be clear: regulation is fine, but it's not enough. Yes, we could regulate industry from the fossil fuel companies downwards. (We could try. I suspect we might hit up against some hard edges if we did, but it's a nice theory.) Yes, we could mandate massed insulation of our homes/offices and give everyone a free air-source heat pump. We'd have to redefine our relationship with money first, but that's pretty much a given anyway. Yes, we could create universal basic services and mandate a Platonic maximum: minimum wage ratio of 20:1. All of these would be good. None of them, not even all of them combined, would be enough – unless they were coupled with Meadows' top leverage point: *Transcend All Paradigms*. I learned of this at Schumacher College in the Masters in Regenerative Economics course. Or rather, I was taught the 12 leverage points and promptly forgot the top one. I spent four years in a state where it seemed not to exist and threw all of my energy at the penultimate one, 'Change the paradigm'. That seemed do-able. I could imagine how it might happen. But 'Transcend All Paradigms'? That means letting go of even the idea that we have to let go of everything. How in the name of all that's good and right and beautiful are we supposed to do that?

I'll try to illustrate how, and bring this chapter towards its conclusion, by telling a story...

A Personal Story

My spiritual path is shamanic. This is to say I live in a world where consensus reality is a tiny part of all actual reality, and the web of life is open to connection by just about anyone who cares enough to put in the time and effort to shed the chains of the death cult and be open to what is. This is a big ask, though. It's huge. Too, there is an argument that 'shed the chains of the death cult' and 'transcend all paradigms' are synonyms. After a fashion. We'll come back to this. On with the story...

Shamanic practice[94] has shaped my life, from veterinary surgeon to novelist, to smallholder, to podcaster, and from practitioner-enquirer to teacher (and still enquirer) of shamanic dreaming. In 2016, after one of those shamanic imperatives which offers no wriggle room at all, I ended up reading for a Masters in Regenerative Economics at Schumacher College, a subject which had not remotely been on my radar before this point.

The learning curve was vertical. We learned classical economics with its inbuilt growth imperative and came to understand why this stood at the heart of our current crisis. Then we created new models, Herman Daly style,[95] so that we could acknowledge the living world as our source and container, with the implicit understanding that, solar radiation apart, we live in a closed system, a fact which orthodox economics chooses consistently to ignore. We learned of planetary boundaries and the speed with which they are being pushed through, to ends we can never know (because complex adaptive systems are complex and adaptive and the one thing we can say for sure about our models is that they're going to be incomplete).

When we'd understood the fragile nature of the outer boundaries, we put a circle inside to make Kate Raworth's Doughnut[96] with the foundation of basic human needs and the understanding that the only safe operating space is one in which our human super-organism becomes regenerative by design, ensuring that the needs of all life are met within the means of the living planet.

With the doughnut as our foundation (it's a torus, for those of you who prefer geometry to empty carbs), we were able to dispense entirely with orthodox models and their toxic blend of cultural narcissism, faux-Darwinism and libertarian pseudo-freedoms.

In their place, we explored Buddhist economics, ecological, feminist, green... and towards the end of first term, someone gave a nod in my direction and mentioned that they knew of a case of shamanic economics in which those about to construct a dam had conducted a ceremony asking the river for permission.

Is your head exploding? Mine did. More than my dawning understanding of the egregious lies on which our economic axioms are based, more than my growing horror at the nature of money as commodified grief, more even than my endless rage at the ways in which orthodox economics denies the sanctity of nature, or even of life... this one passing comment blew all my fuses.

Because endeavouring to placate a river you're about to destroy is not shamanic economics. It's predatory capitalism with a pale green dusting. Economics with a shamanic focus would entail spending weeks, months or (probably) years in the company of the river in question until we knew it – and it knew us – well enough to ask, 'What do you want of me?', in ways that could engage with a serious answer and I seriously doubt if a million tonnes of concrete would come into it. It probably wouldn't make much money, either.

Wrong Question

So I threw away the term paper I'd been planning and instead set out to ask, 'What does shamanic economics look like?' I used all the tools in my kit: I undertook drum-journeys, set dream intents, waking and sleeping... and went home to Shropshire and took my question for long walks in the hills.

My practice at this point was to invite in the deity with whom I work such that it was seeing through my eyes and hearing through my ears and I was seeing/hearing it seeing/hearing through my eyes/ears in a way that changed my perception. Near the end of an hour's walk one Friday morning, I came round a corner and the god stood clear in front of me. So I asked my question and, in one of those rare moments of plain-text communication, was told, 'You're asking the wrong question.'

This was Friday. I had to hand in my term paper by 9am Monday.

Hmmm... 'OK, what's the question?'

And the god said, 'You need to work out what you're here for. Everything else flows from this.'

Which may be obvious when it's said aloud, but nonetheless became my new lode star and has remained so ever since. Unpicking it nudged my partner, Faith, and me toward setting up Accidental Gods membership and podcast and later to establish the Thrutopia Masterclass. More recently, I wrote *Any Human Power*,[97] a Thrutopian political thriller which builds on the ideas presented in these chapters and all that surrounds them.

In the process, it rapidly became clear the answer was, 'To connect with the web of life with sufficient integrity and authenticity that we can ask What do you want of me?, hear the answers in a way that feels grounded and genuine, and respond in real-time.'

Which is fine, if we only knew how to do this – and more importantly, knew how to help other people to do so in large numbers in a useful time frame.

Exploring this last, I spoke on the podcast with Alnoor Ladha[98] and Lynn Murphy, authors of *Post Capitalist Philanthropy: Healing wealth in the time of collapse*.[99] Towards the end, we touched on the concept of Initiation Cultures compared to Trauma Cultures, a topic which Alnoor had opened up in a conversation with the author and trauma therapist, Francis Weller, published in *Kosmos* magazine.[100]

I highly recommend you read the original, but to condense the concept to its barest bones, Weller has come to understand that in healthy human cultures (which here in the UK, would be those that existed before the brutal Roman colonisation of the first century of the Common Era) there are distinct initiatory processes in which an individual undergoes a 'contained encounter with death'. The key is the containment of what might well otherwise be a traumatic event. But the initiate is held by the land, by the elders, ancestors and community and by the ritual itself, the structure of which has evolved out of an understanding of time and place. The end result is that the individual undergoes a radical alteration in their sense of self in a way that opens up their perception of reality and connection to the wider animate earth. They are brought, in fact, to a moment in which they transcend all the earlier paradigms and grow into deeper connection with the web of life.

Trauma cultures, by Weller's contrast, create an almost infinite number of random *uncontained* encounters with death in which there is no holding, no ritual, no elders, no sense of a community holding or connection to the wider web. Thus, when we find ourselves assailed by things that stretch our capacity to cope, our psyche's response is often to shut down, build walls, seek self-created containment and endeavour to find healing in our isolation. This does not – cannot – work, and so we end up more deeply severed from the web of life: insular, isolated and alone.

Clearly this doesn't happen every time, but it's the norm, and the lack of initiation is one of the key indicators of our Western, Educated, Industrial, Rich and (notionally) Democratic – WEIRD – culture.

I am not wholly comfortable with label of trauma, although it's easy to see how it arose – Weller worked closely with Malidoma Somé to gain an understanding of healthy human (initiation) cultures and in our culture, he was a trauma therapist, so the two poles of this particular story were alive for him. I've been considering options and am leaning towards Connected Culture and Disconnected Culture as being less obviously traumatic, but they do lack a certain punch. Labels aside, it seems to me that another way of framing our answer to the god's question is, 'We are here in this time and this place to find a collective way forward to a new style of initiation culture that will offer us resources in the twenty-first century.'

And still, this begs the question: How?

How are we going to step into initiation, to become the best that we can, with the deepest connection to the Web of Life? Particularly given that most of us don't live in a culture which thrives with and through ceremony and ritual, replete with elders who are themselves intimately connected with the web and exist as nodes of consciousness in its awareness.

How, then, do we create our own initiations?

I was sitting with this when, serendipity scattered its star dust and Nina Simons, co-founder of Bioneers,[101] was a guest on the podcast.[102] Nina is an extraordinary elder, with a clarity-of-spirit that shines through all she writes and says. She speaks movingly of a ceremony conducted for her Bioneers group by the Peruvian elder, Oscar Miro-Quesada. At the end of eight hours of ritual he said the following:

"Consciousness creates Matter,

Language creates Reality,

Ritual creates Relationship"

This is quoted in Nina's heart-full book *Nature, Culture and the Sacred*[103] and she spoke of it again on the podcast, reminding us that we can make our own rituals, and giving examples from her own life where she builds rituals regularly. Discussing one where she addressed her own body dysmorphia with a daily post-shower ritual involving scented oils and music, she pointed out that after around six weeks of doing it daily, her inner judge was at its height, telling her she was wasting her time. By around twelve weeks, though, she began to experience real change. She went on to talk of a time when an elder of her land had suggested she go out for a (long) walk holding only the question 'What wants to come through me?' and then write the answer. Her book is largely the product of these rituals.

All of which brings me to this: we have some whole, healthy human cultures still in this world and there are elders and deeply wise individuals from amongst them who are prepared to help us find our way.

Additionally, we have elders amongst us of our own trauma culture who are working daily with their own healing, and that of others, creating rituals that build relationships with ourselves, each other and the living web of life.

In Accidental Gods, we're on the cusp of launching a new set of meditations/rituals that aim to address this and we are not alone: across the world, entire movements are growing under the radar, working with sociocracy and holocracy and the full panoply of social technologies that are so prevalent in *Permaculture* magazine and the circles of life that it touches.

What would happen if we all stepped into lives devoted to, and encompassed by, ritual? What would it feel like? What would the world become?

I have no idea of the answer: we're right at the edge of inter-becoming that Indy Johar speaks of with such integrity.[104] This is the place where emergence happens. It's the place where the levers of change stop being ways to nudge the super-tanker of our existing system five degrees off course, and become instead, the transformation of the super-tanker into something we can barely imagine.

But we need to grasp our courage and take this leap of faith swiftly. For those of us who have grown in the heart of the death cult, it's hard to imagine living in a culture where everyone is interconnected. So much of our history is of people who didn't fit in being shunned, excluded or worse; and most of us don't feel like we really fit in. We're born expecting to be welcomed into a network of connection that helps us become the best of ourselves, and instead we're met with socialisation into separation that teaches us that judgement and its fallout are the natural way of life.

Be that as it may, I don't think we'll get through the bottleneck of the meta-crisis if we don't find our way forward into connection. So I'd like to invite all of you reading this to pour all of your creativity into imagining a world that is fully connected, where we are not born to pay bills and then die, where we can each craft rituals for ourselves and those with whom we engage in our communities of place and passion and purpose, reweaving our capacity to hold ourselves in contained encounters with death so that trauma doesn't dissociate us further, but becomes a route through to growth.

Because this is how we discover the things at which we excel and learn to do them as well as we possibly can, in ways that only we could do, in service to the greater web of life. If we stop to think too hard about it, this could be terrifying. But if we just do it? The world would change overnight.

Let's go, you and me. Out beyond the boundaries of our fears lies a land of unimaginable promise. Let's meet each other there.

'...And it's a human need to be told stories. The more we're governed by idiots and have no control over our destinies, the more we need to tell stories to each other about who we are, why we are, where we come from, and what might be possible.'

Alan Rickman

To help make what is discussed in this book concretely imaginable by the reader, and to move this book towards its end, we turn to a work of fiction, a play. In the short chapter that follows we discuss In It Together *by Kate Webster. We have tried not to include any major spoilers, but you might want to watch the play before returning here to consider our interpretation of it.*

This book has no 'Conclusion'. To have one would make little sense; for TrAd is just getting started; it will surely scale up immeasurably in the coming decade. But this chapter – in a dance with the play, and then closing the book with the TrAd Declaration – is as close as we come to a conclusion. Thoroughly in process, like any true Thrutopia. Work, in progress.

In It Together *by Kate Webster, commissioned as part of the cultural dimension of COP26, can be watched online.*[105]

Chapter 11

To TrAd or not to TrAd?

Morgan Phillips

I first met Rupert Read at a Common Cause get-together in Machynlleth, mid Wales, in 2010. We have kept in touch fleetingly since, mostly online, usually Twitter. I bumped into him in person for the first time in years in 2019 in Oxford Circus. We were both taking in the sight of a parked-up pink boat flanked by hundreds of extinction rebels who – sat in the road – were blocking the Regent Street/Oxford Street intersection.

Extinction Rebellion was springing into life. It was an energising moment, arguably historic, we hugged. Rupert went on to become one of the faces of Extinction Rebellion, inspiring thousands of activists with his speeches, TV appearances, and writing. I don't remember much of what we discussed that day; from my end, it would have been along the lines of 'Wow! Is this really happening?'. But it led to us chatting more over the next few months and years. We've done events together since, and shared thoughts and ideas on the criticality of the times we are living through. This book has been our first formal collaboration.

We have presented here a connected collection of writings on Transformative Adaptation. These essays and articles are not the definitive last word on TrAd. TrAd is an emerging idea/philosophy/practice, it is still gestating. These are the early writings, and so this book is an attempt to capture the *first word* on TrAd.

Rupert and I met up in King's Cross, London, a couple of times in 2022 to plan this book. On our second meeting, he told me about Kate Webster's play; she had written it after reading *This Civilisation Is Finished*, Rupert's brilliant essay-turned-book. We started to play around with the idea of including the play in this book in some way. In the end we decided against it (a play is rarely adequately represented by words on a page), but we encourage you to take 30 minutes to watch it; it communicates the moment Western/globalised civilisation has found itself in, and how we might even yet go forward from that moment.[106]

In It Together, Kate's play,[103] gives us a brief glimpse into everyday life for two women, Gia (in her 30s), and Ruby (in her 60s), in a post 'transformative re-set' [scene 1], or post 'collapse' [scene 2], England. It greatly helps, we believe, to co-imagine the kind of futures that are now possible. Imagining them may be a prerequisite to choosing between them; to determining and shaping a transformed society in which we are coping with the coming impacts.

There are those within the environmental movement and beyond, who will dismiss the likelihood of scene 1 of Kate's play ever coming to pass, let alone scene 2. But for those who are reconciled with the reality of how advanced the climate and ecological crisis now is, the play will resonate. It will hit home for those who have done what Rupert asks us to – accept that **this** *civilisation is finished.*

Through both scenarios, Kate invites us to imagine what life outside of Ruby's modest and threadbare home must now be like for 'ordinary' people. Clearly the second future, post 'collapse', sounds desperate and entirely undesirable; not only has today's civilisation come to an end, but it has been replaced by something very fragile and hazardous. The personal cost of helping others has become unbearable, our planetary support systems are spiralling perpetually downwards, communities and people are breaking down. Neither Rupert nor I, nor the emerging TrAd movement wish to linger *too* long in this imaginative space because, whilst warnings of disaster can have a motivating effect, they can also breed doom, despair, and inaction. They serve as salutary warnings of where we are now headed, on a default setting; but they need to be complemented by alternative visions of where we *could* head instead.

The first future, which is the result of an imagined mid-2020s 'transformative re-set', has similarities with the post 'collapse' scenario in that lives and communities are fragile, and have become far less materially abundant. Crucially though, this future was a chosen one; it was the result of a decision to enact a transformative change. The shock of the big reset has been weathered, the resulting privations and sacrifice have been normalised, and a sense that it will all lead eventually to something more positive is growing. So, whilst many of the trappings (good and bad) of today's civilisation have been abandoned or lost, hope and an optimism of sorts still exists for Ruby and Gia. Progress (newly redefined) feels possible, lives are improving, definitions of the 'good life' have changed. This is a far more *TrAd-like* imaginative space to occupy.

Works of fiction, plays especially, can have multiple powerful effects. They can engage readers and audiences in emotional ways, stirring them to empathise with heroes and anti-heroes; they enable life to be seen through the eyes of others. This of course is why Manda Scott has been taking such trouble to 'theorise' – and write – Thrutopia(s).

And there are at least two ways in which Transformative Adaptation is relevant to the play. The first, I've already commented on: the opening scene, the first scenario, *is* one of Transformative Adaptation having been, and continuing to be, enacted. In this way, it is a believable scenario of active hope, a 'North Star'. The second way is less obvious, but even more telling: the closing scene, the second scenario, is one of a failure to enact TrAd, and more generally of a failure to transform. Kate's brace of Readian scenarios was written before and for COP26. The first scene recounts a (counterfactual) history of activism at the end of COP26 that transformed history. Sadly, of course, we (now) know that this didn't actually happen.

The second scene depicts a future in which indeed it didn't happen. This is our future: unless we manage to change it. So, the second way in which Kate's play is 'about' TrAd is by way of absence: the absence of TrAd in the second scenario should wake us up to and impel us into the deep importance of undertaking TrAd, now: in the absence, very largely, of leadership from states/governments, and in the cold awareness of the failing tide of history to date. In other words, our best option now to stop scene two from coming true, given that there has been no big reset, neither starting at COP26 nor elsewhere, is to undertake Transformative Adaptation. We are *in* the age of consequences; the age in which attempting to adapt is inevitable. Given that COP26 has not put us on the right road (nor indeed COP27 nor 28), we are going to have to improvise a road for ourselves.

Rupert and I encourage you to watch Kate's play especially for these reasons. We hope you watch it, maybe even perform it. Engagement with *In it Together* will hopefully support you to imagine, envision, and ultimately create the world outside of Ruby's window.

We hope too that, with the added support and inspiration of the first words on Transformative Adaptation that you have read, the future or futures you co-imagine and co-create, while under duress, are positive and indeed beautiful.[107]

To close this first word on TrAd, we have collaborated with founder members of the movement to pen a declaration. We strongly welcome anyone who shares this vision for Transformative Adaptation to share the vision on.

Transformative Adaptation: through words, through deeds, through all.

Transformative Adaptation: A Declaration

The TrAd Collective

TrAd is a practical philosophy.

It is too late to head off disasters. They are now coming; they are here. It is too late to prevent dangerous human-caused climate change or much ecocide. It is therefore time to start taking adaptation seriously.

We Declare That...

We will begin to adapt ourselves and our communities toward climatic degradation in a way that mitigates global environmental destruction.

We will work with nature rather than against nature, and seek in the process to transform society in the direction it needs to change.

We call too for deep adaptation: preparing for potential societal collapse.

Transformative (and deep) adaptation will be modelled in our actions and plans.

We assume that government will not take the initiative, but we will nevertheless challenge them to do so through actions which model it.

We are not a political party and note that the existing 'democracy' has failed us. We need real democracy to be revived.

We believe in Citizens Assemblies. We cannot rely on government to set these up; we will seek to model them ourselves at every level (locally, nationally, globally), and challenge Governments to adopt them and their outcomes.

We seek a diverse and inclusive society. We seek unity. We advocate a 'co-liberation' model for advancing this aim. We determine to work together to overcome oppressions and prejudice and to free ourselves together.

Our actions are the strongest way we tell a story, and change the narrative.

Our actions will be intelligibly and visibly articulated as senseful acts of beauty. We do not believe in actions that seem to threaten, or senselessly destroy. Our actions will always be meaningful.

We intend system-change, from the bottom up.

We intend to create a better world, a more loving world, or at least a more survivable world.

… We commit to acting from here on in the spirit of this declaration…

Endnotes

1. The three pillars are explicated at some length in later chapters.
2. In this endeavour, we stand on the shoulder of previous pathfinding efforts, very notably that of my co-editor, Morgan Phillips, in his brilliant and groundbreaking book *Great Adaptations: In the shadow of a climate crisis*, Arkbound: Bristol, 2021.
3. See Chapter 7 for detail on this.
4. This is not an emergency... it's much more serious than that; emerge: www.whatisemerging.com/opinions/climate-this-is-not-an-emergency-it-s-much-more-serious-than-that Read, R. and Knorr, W. 2022.
5. *Deep Adaptation*; edited by Jem Bendell and Rupert Read. Polity: London, 2021. See also www.politybooks.com/blog-detail/deep-crisis-deep-adaptation
6. https://climatemajorityproject.com/expert-statement
7. For a strong up to the present presentation, see this interview of Johan Rockstrom by Kevin Anderson: www.youtube.com/watch?v=ILq8e73-FAw
8. For discussion, see:
   Iain McGilchrist's oeuvre,
   Daniel Schmachtenberger's thought: www.civilizationemerging.com,
   Hospicing Modernity; Vanessa Andreotti. North Atlantic Books: New York, 2021,
   This Civilisation is Finished; Rupert Read and Samuel Alexander. Simplicity Press: Melbourne, 2019.
9. For the disastrous effects of rampant inequality, the prime source-reference remains the work of Wilkinson and Pickett. For a vision of the economy which goes beyond growth while conducing to the needs of people, the greatest source is the late Herman Daly, while Kate Raworth and Tim Jackson also very worth consulting.
10. On which, in the current context, see especially, 'Where value resides', *Environmental Ethics*, 37:3, 2015, pp.321-340; Tom Greaves and Rupert Read. See also www.aru.ac.uk/global-sustainability-institute-gsi/research/ecosystems-and-human-wellbeing/debating-natures-value and www.aru.ac.uk/global-sustainability-institute-gsi/research/ecosystems-and-human-wellbeing/debating-natures-value/resources, and *Debating Nature's Value: The concept of natural capital*; edited by Victor Anderson, Berlin: Springer, 2018.
11. UEA Publishing Project: Norwich, 2021.
12. Canbury Press: Kingston upon Thames, 2018.
13. On this point, see my 'Existential investigations of our existential crisis', co-authored with Joe Eastoe, *Think*, Volume 22, Issue 65, Autumn 2023, pp.65-71.

14 A key part of the thinking and practice underscoring this practice can be found in the life's work of my teacher Joanna Macy. And see www.abc.net.au/religion/rupert-read-climate-grief-could-be-the-making-of-us/14076522

15 For full-scale amplification of it, see: *A Film-Philosophy of Ecology and Enlightenment*; Read, R. Routledge: London, 2019.

16 The work of John Foster is important to this. See especially: *Realism and the Climate Crisis*; Foster, J. Bristol University Press: Bristol, 2022.

17 For discussion, see *Why Climate Breakdown Matters*; Rupert, R,. Bloomsbury Press: London, 2022.

18 This final paragraph is of course influenced by Antonio Machado's poetry.

19 *Facing up to Climate Reality*; Foster, J. eds., Greenhouse Think Tank, 2019.

20 'The Great Turning'; Joanna Macy. www.ecoliteracy.org/article/great-turning

21 Bioregional, One Planet Living: www.bioregional.com/one-planet-living

22 https://makerojavagreenagain.org

23 *Great Adaptations: In the Shadow of a Climate Crisis*; Morgan Phillips, Arkbound: UK, 2021.

24 *Climate Adaptation: Accounts of Resilience, Self-Sufficiency and Systems Change*; Morgan Phillips, Arkbound: UK, 2021.

25 See www.climateemergency.co.uk

26 https://climateemergencycentre.co.uk

27 www.ClimateMajorityProject.Earth

28 www.thetimes.co.uk/article/cop26-world-set-to-heat-up-by-2-7c-says-analysis-of-net-zero-pledges-8mkzd0vsc

29 www.politybooks.com/blog-detail/deep-crisis-deep-adaptation

30 https://transformative-adaptation.com

31 www.griefyork.com

32 www.ecologicalcitizen.net/pdfs/v03n2-10.pdf

33 www.theosthinktank.co.uk/comment/2022/01/19/rupert-read-on-nonviolence-the-climate-crisis-and-the-power-of-emotions

34 https://systems-souls-society.com/what-next-on-climate-the-need-for-a-moderate-flank

35 www.youtube.com/watch?v=eVhpcpJcNkQ

36 I don't dwell on the crucial potential of workplaces (and professions) in this chapter. For discussion of it, see my 'Between XR and CoP: Pivoting climate movement strategy from the radical flank effect to a 'moderate' flank, via a shift toward workplace-based activism', www.greenhousethinktank.org/between-xr-and-cop-pivoting-climate-movement-strategy-from-the-radical-flank-effect-to-a-moderate-flank-via-a-shift-toward-workplace-based-activism

37 www.wildcard.land

38 https://climateemergencycentre.co.uk
39 www.trustthepeople.earth
40 This final section is influenced by the works of the late great Bruno Latour, who I was fortunate enough to encounter in person and learn from back when I was a graduate student.
41 'World will miss 1.5C warming limit – top UK expert': Stallard, E. and Rowlatt, J., 2023. www.bbc.co.uk/news/science-environment-66256101
42 'We're calling on people to get involved in climate community action', Climate Majority Project, 2023.
www.youtube.com/watch?v=UHuKxqeVNPM&feature=youtu.be
43 *Operating Manual for Spaceship Earth*; Fuller, B. Simon and Schuster: USA, 1970
44 Speech to United Nations General Assembly (Global Environment): Thatcher, M., 1989.
www.margaretthatcher.org/document/107817
45 'Climate change: 12 years to save the planet? Make that 18 months': McGrath, M., 2019. www.bbc.co.uk/news/science-environment-48964736
46 'Climate Breakdown Has Begun with Hottest Summer on Record, Secretary-General Warns, Calling on Leaders to 'Turn Up the Heat Now' for Climate Solutions': Guterres, A., 2023.
https://press.un.org/en/2023/sgsm21926.doc.htm
47 *Are you scared yet*? Thompson, J., 2023. www.eastsuffolk.gov.uk/assets/Environment/Green-Issues/Events/Joolz-Thompson-iFarm-Sustainability-v3.pdf
48 *Great Adaptations: In the Shadow of a Climate Crisis*; Phillips, M. Arkbound: UK, 2021.
49 As the world burns you have a choice: Climate Museum UK, 2018.
https://climatemuseumuk.org/2018/07/25/as-the-world-burns-you-have-a-choice
50 Community Climate Action: Climate Majority Project, 2023.
https://climatemajorityproject.com/community-climate-action
51 'UN: only small Farmers and Agroecology can feed the World': Ahmed, N., 2014. www.tni.org/en/article/un-only-small-farmers-and-agroecology-can-feed-the-world
52 Suffolk's farming and wildlife advisory group: https://suffolkfwag.co.uk
53 Project Drawdown: www.drawdown.org
54 Perceptions Matter: Common Cause Foundation, 2016.
https://commoncausefoundation.org/wp-content/uploads/2021/10/CCF_survey_perceptions_matter_full_report.pdf
55 Humankind – A hopeful history: Bregman, R., Bloomsbury: UK, 2020.
56 *Citizens: Why the Key to Fixing Everything is All of Us*; Alexander, J., Canonbury Press: UK, 2022.

57 *Together: A Manifesto Against the Heartless World*; Temelkuran, E., Harper Collins Publishers: UK, 2022.
58 The Dirt Is Good Project: www.dirtisgoodproject.com
59 https://accidentalgods.life/exploding-the-myth-of-a-farm-free-future-and-fake-meat-with-chris-smaje
60 www.theguardian.com/books/2014/nov/20/ursula-k-le-guin-national-book-awards-speech
61 https://journal.workthatreconnects.org/2022/03/24/miki-kashtans-work-of-reconnecting
62 https://podcasts.apple.com/gb/podcast/the-great-simplification-with-nate-hagens/id1604218333?i=1000595204874
63 www.robhopkins.net/2020/06/15/when-the-chickens-of-imagination-come-home-to-roost
64 https://davidgraeber.org/books/bullshit-jobs
65 www.independent.co.uk/news/uk/politics/ukraine-fuel-bills-martin-lewis-b2032796.html
66 www.youtube.com/watch?v=6iM6M_7wBMc
67 www.huffingtonpost.co.uk/rupert-read/thrutopia-why-neither-dys_b_18372090.html
68 https://every.to/p/breaching-the-trust-thermocline-is-the-biggest-hidden-risk-in-business
69 www.youtube.com/watch?v=_s30m6Bpj2U
70 https://uk.bookshop.org/p/books/the-dawn-of-everything-a-new-history-of-humanity-david-graeber/5715204?aid=7671&ean=9780141991061
71 www.wired.co.uk/article/taiwan-sunflower-revolution-audrey-tang-g0v
72 www.flatpackdemocracy.co.uk
73 https://towardsdatascience.com/what-is-quadratic-voting-4f81805d5a06
74 https://medium.com/giveth/conviction-voting-a-novel-continuous-decision-making-alternative-to-governance-aa746cfb9475
75 www.furtherfield.org/culturestake
76 https://moreequalanimals.com/posts/book-launch
77 https://wearenotdivided.reasonstobecheerful.world/taiwan-g0v-hackers-technology-digital-democracy
78 www.kateraworth.com/doughnut
79 www.youthxyouth.com
80 *Debt: The First 5000 Years*; David Graeber, Melville House Publishing: London, 2014. https://uk.bookshop.org/p/books/debt-the-first-5000-years-david-graeber/2823513?ean=9781612194196
81 https://onezero.medium.com/survival-of-the-richest-9ef6cddd0cc1

82 Learn more on Nate Hagens' podcast: https://podcasts.apple.com/gb/podcast/the-great-simplification-with-nate-hagens/id1604218333?i=1000562226501

83 www.riversimple.com/governance

84 www.nobelprize.org/prizes/chemistry/1977/prigogine/facts

85 www.nature.com/articles/424030a

86 https://exploriter.com/concepts/phases-of-tragedy

87 https://youtu.be/ljOQB608ylQ?si=qg8SoNH_kOWjeb4N

88 https://islandsofcoherence.net

89 www.traditionaldreamfactory.com

90 www.stockholmresilience.org/research/planetary-boundaries.html

91 *Metabolical: The Truth about Processed Food and how it Poisons People and the Planet*; Dr Robert Lustig, Hachette: London, 2021.

92 https://chrissmaje.com/blog

93 www.donellameadows.org/wp-content/userfiles/Leverage_Points.pdf

94 I would argue quite fiercely that nobody with a western education / socialisation can ever be a shaman. But nonetheless, the path of shamanic practice is incredibly enlightening and enlivening. I am also aware that this is a fairly contentious viewpoint and you're free to believe differently.

95 https://theconversation.com/the-inconvenient-truth-of-herman-daly-there-is-no-economy-without-environment-193848

96 https://doughnuteconomics.org/about-doughnut-economics

97 https://septemberpublishing.org/product/any-human-power-hb

98 See Alnoor's articles in *Permaculture* magazine. 'Seeing Weitko' (Issue 105) and 'The Poverty of Progress' (Issue 106)

99 www.postcapitalistphilanthropy.org

100 www.kosmosjournal.org/kj_article/deschooling-dialogues-on-initiation-trauma-and-ritual-with-francis-weller

101 https://bioneers.org

102 https://accidentalgods.life/on-nature-culture-and-the-sacred-with-elder-and-visionary-nina-simons

103 www.ninasimons.com/writing

104 https://accidentalgods.life/becoming-intentional-gods-claiming-the-future-with-indy-johar-of-the-dark-matter-labs

105 www.youtube.com/watch?v=ifFC_uTR9PA

106 *In It Together / Good CoP, Bad CoP*: Webster, K. 2021. www.youtube.com/watch?v=ifFC_uTR9PA

107 Thanks to Rupert Read for input into this chapter.

Enjoy this book?

Permanent Publications is a small permaculture enterprise and ordering your books direct is like shopping locally.

Tell your friends! Your positive recommendations hugely help us reach a world in desperate need of positive and practical solutions.

We publish a range of books to empower and inspire change-makers the world over, from no dig organic growing, food forests and permaculture, to natural building, renewable technology and connecting with nature.

If you enjoyed *Transformative Adaptation*, why not try these titles related to transforming culture.

THE EARTH CARE MANUAL
Patrick Whitefield

CULTURAL EMERGENCE
Looby Macnamara

PERMACULTURE DESIGN
Aranya

For the full list of our solution-based titles visit
www.permanentpublications.co.uk

Our books are also available in:

North America: https://tinyurl.com/ChelseaGreen

Australia: https://tinyurl.com/Peribo

Permanent Publications also publishes *Permaculture* magazine

Permaculture Solutions

We publish *Permaculture* magazine – the voice of changemakers.

I really do encourage you to cancel any subscriptions you may have to the mainstream media... And instead you could use that money to subscribe to Permaculture magazine. It is a genuinely regenerative, huge hearted, strong hearted chronicle of good things that are happening now. It's beautifully designed and the articles are genuinely inspiring.

Manda Scott,
Author of bestselling *Boudica* series,
and host of Accidental Gods podcast

A subscription to *Permaculture* magazine includes:

- **FREE** digital access to 30+ years of searchable back issues
- A reusable **25% discount code** for ALL our Permanent Publications titles
- Subscriber-only offers
- A **20+%** discount on the cover price
- A quarterly dose of solutions delivered straight to your home

Digital only
Paper and postage free
Just £13.99

Print with FREE Digital
1 or 2 year subscription options

Direct Debit
Print with FREE Digital
Price locked until 2030

Or visit:
www.permaculture.co.uk/subscribe